Margaret C. Harrell Harry J. Thie Peter Schirmer Kevin Brancato

Aligning the Stars

Improvements to General and Flag Officer Management

T0159313

Prepared for the

Office of the Secretary of Defense

Approved for Public Release; Distribution Unlimited

RAND

National Defense Research Institute

The research described in this report was sponsored by the Office of the Secretary of Defense (OSD). The research was conducted in the RAND National Defense Research Institute, a federally funded research and development center supported by the OSD, the Joint Staff, the unified commands, and the defense agencies under Contract DASW01-01-C-0004.

Library of Congress Cataloging-in-Publication Data

Aligning the stars : improvements to general and flag officer management / Margaret C.
    Harrell ... [et al.].
        p. cm.
    Includes bibliographical references.
    "MR-1712."
    ISBN 0-8330-3501-0 (pbk. : alk. paper)
    1. United States—Armed Forces—Officers. 2. Generals—United States. 3.
Admirals—United States. 4. United States—Armed Forces—Personnel management.
I. Harrell, Margaret C.

UB412.A4 A795 2004
355.3'31'0973—dc22

                                                                    2003024739

The RAND Corporation is a nonprofit research organization providing objective analysis and effective solutions that address the challenges facing the public and private sectors around the world. RAND's publications do not necessarily reflect the opinions of its research clients and sponsors.

RAND® is a registered trademark.

Cover design by Stephen Bloodsworth

© Copyright 2004 RAND Corporation

Published 2004 by the RAND Corporation
1700 Main Street, P.O. Box 2138, Santa Monica, CA 90407-2138
1200 South Hayes Street, Arlington, VA 22202-5050
201 North Craig Street, Suite 202, Pittsburgh, PA 15213-1516
RAND URL: http://www.rand.org/
To order RAND documents or to obtain additional information, contact
Distribution Services: Telephone: (310) 451-7002;
Fax: (310) 451-6915; Email: order@rand.org

Career patterns of general and flag officers (G/FOs) are of interest to Congress, the Secretary of Defense, and the military services. For example, the House and Senate conferees for the National Defense Authorization Act for Fiscal Year 1997 stated in their report that "the current general and flag officer selection, assignment, and development process may not effectively contribute to the preparation of those officers for increasing levels of responsibility and maximum performance efficiency at each level of assignment." Among specific stated concerns were

> the tempo with which general and flag officers are rotated through important positions; the effect of this tempo both on the effectiveness of individual officers in each position to which they are assigned and on the overall value these officers add in each position to which they are assigned; and the consequences of requiring general and flag offices to retire upon completion of 35 years of service.[1]

The Secretary of Defense has expressed similar concerns:

> I kept noticing that people that were in their jobs 6, 8, 10, 12, 13, 15 months. And general officers, flags. I know that if you had a need to punch a ticket to get your schooling, your training, to get your joint pieces under Goldwater/Nichols, there is tremendous pressure to do that. I also know that it's difficult for people to really learn a job

---

[1]National Defense Authorization Act for Fiscal Year 1997; Conference Report to Accompany H.R. 3230 (1996).

and then do it well enough and know what their mistakes were because you have to be around long enough to see some of it.[2]

In the military services, the concern is to maintain promotion opportunity throughout a hierarchy of 10 grades through which officers can flow. This promotion flow, especially for the middle management grades of O-4 through O-6, was carefully crafted as part of the Defense Officer Personnel Management Act of 1980. Whether or not they are the best flow rates can be argued, but expectations have been set for about 20 years based on them. The concern is that longer service in a particular grade will clog promotion flow at lower grades.

What are the appropriate practices for assigning and developing G/FOs? What are the effects of changing them? This report addresses these questions by examining empirically current patterns of G/FOs, by examining how private-sector executives are assigned and developed, by reviewing the literature of career management and executive development, and by analyzing how changed assignment and development practices might affect promotion probability and service tenure.

This report should interest the manpower and personnel policy and analytical communities as well as military officers and defense policymakers. This research project was sponsored by the Director for Officer and Enlisted Personnel Management in the Office of the Under Secretary of Defense (Personnel and Readiness).

The research was conducted for the Office of the Secretary of Defense within the Forces and Resources Policy Center of the RAND National Defense Research Institute, a federally funded research and development center sponsored by the Office of the Secretary of Defense, the Joint Staff, the unified commands, and the defense agencies. The principal investigators are Harry Thie and Margaret Harrell. Comments are welcome and may be addressed to harry_thie@rand.org or margaret_harrell@rand.org. For more information on the Forces and Resources Policy Center, contact Director Susan Everingham, susan_everingham@rand.org, 310-393-0411, extension 7654.

---

[2]Donald Rumsfeld, as quoted in the *Washington Post*, July 22, 2001.

CONTENTS

FIGURES

TABLES

BACKGROUND

Members of Congress and senior members of the Department of Defense (DoD) worry that general and flag officers (G/FOs) change jobs too frequently and, consequently, do not spend enough time in an assignment to be as effective as they could be, develop the skills they need for subsequent assignments, or remain long enough to be accountable for their actions. Furthermore, these decisionmakers are concerned that the careers of the most-senior officers do not last long enough. For their part, the military services concern themselves with the flow of promotions through 10 officer ranks, O-1 through O-10.[3] This flow, particularly for the more senior officers, was carefully crafted as part of the Defense Officer Personnel Management Act of 1980, and, whether or not the best policy, it has conditioned officer expectations for more than two decades. The concern is that lengthening the tenure of senior officers could clog the system, causing promotions to stagnate throughout the officer corps.

THIS STUDY

What, then, are the appropriate practices for assigning and developing G/FOs? If current practices change, what would the effect be? This study attempts to answer these questions. It does so by first developing an empirical picture of how the current system manages

---

[3]See, for example, the "Officer Flow Management Plan" in DoD (2000).

G/FOs, reviewing the literature about the private sector to determine how organizations in it manage their senior executives, and modeling different ways of managing the most-senior military officers. The modeling goal was to identify management approaches that addressed the concerns described above and identified the effects of implementing them.

WHAT THE CURRENT SYSTEM LOOKS LIKE

There are about 900 G/FOs in DoD. About 50 percent are O-7s, about 35 percent O-8s, and about 15 percent O-9s and O-10s. Although G/FOs have different career fields, this study focuses on the line category, the one directly associated with the conduct of warfare.[4] Officers in this category typically command large combat formations in the services or serve as combatant commanders. Occasionally, a line officer will serve in another field, such as technical and support; conversely, with less frequency, those in other career fields may have line assignments.

Most G/FO assignments last less than 30 months. Officers who reach the highest rank typically have two assignments as an O-7 and one in each rank thereafter. Promotion tends to occur quickly. While officers spend three years as O-7s, they spend about two to two-and-a-half years as O-8s and two-and-a-half years as O-9s. Most O-10s retire with about 33–35 years of commissioned service, having served less than 10 years as a G/FO. Other G/FOs who retire have similar amounts of service because those promoted to O-10 typically have been advanced to O-7 at an earlier point in their careers than most new flag officers.

The key aspect of this study is the distinction between what we call "developing" jobs and "using" jobs. This distinction rests on the principle that work experience accumulates through a variety of

---

[4]Our sponsor asked us to focus on positions for line officers because they are the officers historically promoted to O-10. Based on the empirical data, we included in our analysis armor, infantry, and field artillery officers in the Army; unrestricted line officers in the Navy; pilots and navigators in the Air Force; and line officers in the Marine Corps. Once we had this subset of officers, we included in our analysis all the assignments that officers in these specialties had as G/FOs, which contained some assignments to technical, support, and, in a few instances, even professional positions.

manager and executive assignments that prepare the individual for increasingly demanding and complex jobs. Early assignments build functional skills, organizational knowledge, and personal insights. Later jobs tend to have more complex and ambiguous responsibilities that draw on the skills and knowledge developed in earlier assignments. Thus, some jobs develop an individual's skills, while others use skills previously developed. We conclude that using jobs should be longer than developing jobs, and our research into literature about the private sector supports this conclusion.

In devising different management approaches for flag officers, we assumed that all O-7 jobs are developing jobs and all O-10 jobs are using jobs. As for the jobs in the middle—those at O-8 and O-9—we assumed that the O-8 jobs that appeared frequently on O-9 or O-10 resumes and O-9 jobs that appeared frequently on O-10 resumes are developing jobs. Each service has a number of G/FO jobs that rarely show up on the resumes of O-9s or O-10s. We designate these as low-frequency jobs and not typical of those intended to develop officers for the most-senior assignments. We categorize these as using jobs at the O-8 and O-9 levels. We then identified the jobs at the O-8 and O-9 levels that are never filled by an officer promoted to O-10. These, too, become using jobs because they occur at the end of an officer's career. These rules were designed to be conservative in identifying using jobs because all G/FOs are eligible for promotion or a new assignment, so in theory anything short of Chairman of the Joint Chiefs of Staff could be considered a developing job.

The current system shows little connection between types of jobs and their duration. Assignment lengths in O-8 and O-9 jobs average from 20 to 26 months. Median assignment length of O-10 jobs ranges from 26 to 32 months. Although civilian counterparts tend to become CEOs at about the same age that O-10s get promoted, assignment tenure differs substantially. The average O-10 serves for three-and-a-half years, and almost 90 percent retire voluntarily before reaching age 60. CEOs serve for almost eight-and-a-half years, and less than a third depart before reaching 60; more than half retire in accordance with corporate policies.

MODELING NEW CAREER PATTERNS

The basis of the modeling analysis was a variation in the tenure between developing assignments and using assignments. Developing assignments were shorter than using ones. We used two independent models to explore different management approaches. Outputs included number promoted, promotion probability to each grade, probability of an O-7 reaching O-10, number of officers not promoted, average time in service, average time in grade for those promoted, average time in grade for those retiring, and average time in job.

The best approach that emerged was one in which developing assignments lasted two years and using assignments four because it met the criteria of maximizing stability and accountability without sacrificing promotion opportunity. In most cases, more officers get promoted to O-7 than under the current system. The number promoted to O-8 either equals or exceeds the current system, and the number promoted to O-9 increases for all services except the Army. Promotions to O-10 decrease for all services by about half because the length of time that officers serve as O-10s increases considerably. Average career length increases for all grades except O-7; however, O-7s will serve in assignments longer than they do today. Average assignment length increases for all pay grades in all services.[5]

CAVEATS AND CONCERNS

Although we believe that the research strongly supports the distinction between developing jobs and using jobs, it is important to note, for several reasons, that the categorization presented here is descriptive, not prescriptive. First, while we could observe how officers are developed today, it is not clear that this would be the best way to do it in the future. Second, causality is ambiguous: Do officers with cer-

---

[5]The average assignment length for O-7s will be 24 months, which is more than the current lengths, which range from 17 to 19.7 months. The average assignment length for O-10s will be 48 months, compared to the current average of 25.3 to 32.1 months. The average length of O-8 and O-9 assignments will depend upon the proportion of jobs that the services determine are developing and using. Our analysis indicates an increase in assignment length. In fact, average assignment length will increase even if only 10 percent of O-8 and O-9 assignments are longer using assignments.

tain experience get promoted, or do officers who have a greater chance of promotion get certain assignments? Finally, the services might not categorize jobs the same way we did.

Additionally, during the course of our research, several concerns were raised about repercussions from the proposed management change:

- **Retention.** While we heard concerns that officers would not be willing to serve longer time in service and in longer assignments, our interviews with serving and retired G/FOs suggest that retention will continue to be an individual issue; there will also be voluntary leavers and unexpected retirements, but retention of sufficient numbers of G/FOs should not be a problem. Analytically, we can also assert that if officers do not behave as predicted, the system may not achieve all the increases in stability and accountability—but it will look no worse than today's system.

- **Flexibility.** We agree with assertions that the system must remain flexible and that an improved system should not be overly rule bound; performance and logic are more important.

- **Compensation.** Many of the senior officers we interviewed mentioned the compensation system; existing shortcomings of the compensation system will become even more evident if officers serve for longer careers.

CONCLUSIONS AND RECOMMENDATIONS

Conclusions

With a few exceptions—a chief of service, for example—the current system does not determine assignment length based on the inherent nature of the job or the way the job is used to develop officers. It should. Distinguishing between developing assignments and using assignments will mitigate the concerns of Congress and senior defense officials and do so without congesting the promotion system. The management changes suggested in this report could be implemented largely within the legislative authority of DoD. The Title 10 authority permitting 40-year careers for O-10s and 38 years for O-9s coupled with a mandatory retirement age of 62 generally is sufficient. However, a change in law could give the services more flexibility to implement the management approach described here. Additional changes, such as to the compensation and retirement system, may also be warranted and would require new authority.

Recommendations

The services should categorize their G/FO positions as either using or developing and determine the desired tenure for each. They need to confirm that they are going to continue developing officers using the assignments that they have in the past.[6] Furthermore, some using assignments may need to be shorter than four years, and some developing assignments longer than two. The optimum time in a job should vary by grade, community, and the inherent nature of the duty. Thus, in line with the analysis in this report, developing assignments would be shorter than using ones. In general, we recommend two years of developing and four years of using for line officers, subject to the review described above. Assignments outside the line community may be longer than those in it. Further, we recommend that officers have three developing jobs in their O-7 and O-8 years and one during their O-9 tenure.

This research suggests the implementation of a system that would increase the tenure of senior officers in assignments, which should foster greater stability and accountability. We recognize that any transition to a new system will encounter difficulties. However, we do not anticipate any retention problems. Our research indicates that retention will continue to be an individual issue conditioned by family concerns and other issues.

---

[6]Our research examines the effect of assigning groups of officers to particular positions for various durations to determine effects on promotion and career outcomes. Additional research is needed to examine assignments based on the developmental needs for individual officers to gain required competencies to fill key positions in the future.

ACKNOWLEDGMENTS

We are grateful for the assistance, facilitation, and interaction provided by our sponsoring office, specifically Colonel Jim Wilkinson, Colonel Christine Knighton, and Brad Loo. We are indebted to the many serving and retired general and flag officers who spent time sharing their perspectives of the current system and reacting to possible changes proposed.

We appreciate the cooperation and assistance of the various general officer, flag officer, and senior leader management offices that expressed their views of the current management process. In particular, Colonel Julie Sennewald was helpful in conveying these views.

The Defense Manpower Data Center and the Washington Headquarters Services' Directorate for Information Operations and Reports provided data that greatly aided our analysis.

This report benefited from the assistance and intellectual contributions of many colleagues at the RAND Corporation, including John Boon, Robin Cole, Frank Lacroix, Susan Everingham, Jeff Isaacson, Jerry Sollinger, and our reviewers, Herb Shukiar and Al Robbert. Also, we thank RAND's Stephen Bloodsworth, who designed the cover, and Phillip Wirtz, who edited and formatted the document.

ABBREVIATIONS

CEO	chief executive officer
DIOR	Directorate for Information Operations and Reports (Washington Headquarters Service)
DMDC	Defense Manpower Data Center
DoD	Department of Defense
JDAMIS	Joint Duty Assignment Management Information System
NATO	North Atlantic Treaty Organization

Chapter One

INTRODUCTION

BACKGROUND

The Secretary of Defense has expressed concern that general and flag officer (G/FO)[1] assignments are too short, that the amount of service after promotion is too short, and that their careers do not last long enough. The Secretary is also concerned that the way G/FOs are managed currently causes high turbulence and turnover in assignments, the loss of vigorous and productive officers to retirement from the military, and the retirement of G/FOs without the minimum expected time in their last pay grade. Additionally, the Office of the Secretary of Defense is concerned that such rapid turnover of assignments reduces organizational effectiveness, dilutes individual accountability among the leadership, limits career satisfaction of senior officers, and erodes the confidence of junior and mid-level officers, who see their military leadership moving so quickly through their organizations that they gain no more than a superficial understanding.[2] Like the military, the private sector also develops its senior

---

[1] General officers of the Army, Air Force, or Marine Corps, and flag officers of the Navy include those in pay grades O-7 (i.e., brigadier general, rear admiral [lower half]), O-8 (i.e., major general, rear admiral), O-9 (i.e., lieutenant general, vice admiral), and O-10 (i.e., general, admiral). By law, there are about 900 G/FOs, of which approximately 50 percent are O-7s, 35 percent are O-8s, and 15 percent are O-9s and O-10s.

[2] The Secretary of Defense has stated that he has observed "very rapid changes of assignment . . . numbers are down around 12 months, 14 months, 16 months, 18 months. That's not very long. One of the effects of that is they get into the job, just start learning it, and then it's just about time to say goodbye and they're out of it onto something else. The disadvantage of that is obvious; people don't have enough time

---

executives through job rotations; however, at the highest level, the most senior executives serve, on average, for eight years and retire later, and this longer tenure correlates with higher organizational performance.[3]

OBJECTIVE

The Department of Defense asked RAND's National Defense Research Institute, to assess promotion, assignment, and tenure issues within G/FO management. This research project was designed to establish the baseline assessment of what G/FO careers currently look like, to analyze possible changes to the current management, and to assess whether such changes might address the Secretary's concerns. The research approach included a review of private-sector literature to understand how private-sector organizations manage their senior executives, analysis of promotion patterns and management of G/FOs from 1975 to 2002, modeling and assessment of different career models and the resulting policies and practices, and interviews with senior military officers to capture their understanding of the current system and to comprehend likely behavioral responses to a changed system.

DATA SOURCES FOR THE BASELINE

The Washington Headquarters Services' Directorate of Information Operations and Reports (DIOR) provided the primary G/FO database used in this report. It was used to generate overall historical patterns, to provide detailed information regarding common sequences of jobs, and to generate inputs for modeling. This electronic database is the result of aggregation of the General and Flag Officer Roster, an exhaustive list published monthly by DIOR. It includes all active and reserve G/FOs from selection for promotion to O-7 until retirement. It tracks officers by name, rank, specialty, service, job title, and unit—

---

there to really set goals, put them in place and work them forward. The advantage of it is that individuals get a chance to do a variety of different things and punch a number of different tickets." SECDEF Town Hall Meeting, August 9, 2001 (quote taken from a DoD news transcript).

[3]Lucier, Spiegel, and Schuyt (2002).

all recorded as of the first of the month when such changes as promotion or change of position occur.

Two databases were used to verify and supplement the DIOR
database. The first was the G/FO database maintained at the Defense
Manpower Data Center (DMDC). This database contains the personnel history (starting in September 1975) of all officers promoted
to the rank of O-7 on or after January 1, 1990. Fields include name,
date of birth, service, rank, occupational code, unit identification
code, and unit address. Although the data in the DMDC database
were more complete than those in the DIOR database, the lack of a
job title and unit name made it infeasible to use the DMDC data to
perform the filtering of positions at the center of RAND's analysis.
We found general historical patterns to be consistent with those generated by the DIOR database.

The second supplemental database was the Joint Duty Assignment
Management Information System (JDAMIS), also maintained by
DMDC.

JDAMIS is a relational database, listing extensive data on "joint"
positions and the officers who have served in at least one position; it
was used solely to supplement the previous two databases with an
officer's basic active service date and to verify the start and end date
of an officer's tenure in a position.

The DIOR database, augmented by other sources, was used to categorize positions as either developing or using (discussed more in
Chapter Four). Because there are known inconsistencies in the DIOR
electronic database before 1987, we performed substantial quality-
control efforts—through comparison with DIOR paper records, the
other electronic databases, and official officer biographies.

ORGANIZATION OF THIS REPORT

The next chapter examines G/FO careers in the current system and
quantifies the rate of movement through assignments and pay
grades. Chapter Three provides a theoretical and empirical framework to examine careers. Chapters Four and Five apply the theoretical framework to G/FOs, determine the career model for senior
military officers that best addresses concerns expressed by the Office

of the Secretary of Defense while minimizing the impact on promotion opportunity (the primary service concern), and model the effects of such management changes. Chapter Six discusses potential concerns about management change and provides insights gained during interviews. Chapter Seven contains conclusions and recommendations. The appendices provide more detailed analysis of current G/FO management (Appendix A), a list of management alternative cases modeled (Appendix B), and both tabular (Appendix C) and flow-based (Appendix D) output of the modeled output; Appendix E is a discussion of executive compensation.

Chapter Two

WHAT DO GENERAL AND FLAG OFFICER CAREERS LOOK LIKE IN THE CURRENT MANAGEMENT SYSTEM?

MULTIPLE CAREER PATTERNS

By law, there are about 900 G/FOs in the Department of Defense (DoD), of whom approximately 50 percent are O-7s, about 35 percent are O-8s, and about 15 percent are O-9s and O-10s. In this section, we review the several broad career patterns that exist within military officer management. Figure 2.1 illustrates the discussion below. The arrows in the figure show horizontal and vertical mobility.

There are 10 commissioned officer pay grades, O-1 to O-10. Grades O-7, O-8, O-9, and O-10 constitute the G/FO grades.[1] Officer occupations and skills fall into three broad categories: professional, technical and support, and line. The line community has four kinds of positions that provide or require different kinds of experience and expertise: military/naval skills, service skills and culture, corporate skills and culture, and military experience.

The professional category includes approximately 100 G/FOs who serve in such fields as medical, chaplain, and legal. These career fields are distinct in that officers do not migrate between them. Also,

---

[1]We use pay grade designations in this study. The ranks for O-7 are brigadier general and rear admiral (lower half); for O-8, major general and rear admiral (upper half); for O-9, lieutenant general and vice admiral; and for O-10, general and admiral.

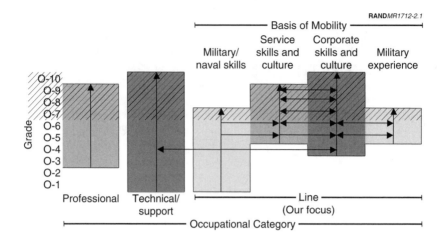

Figure 2.1—Multiple Career Patterns Exist for Officers

they typically begin at grade O-3. Officers enter at a higher grade because of either educational attainment or prior experience. Most of these occupations end at grade O-7 or O-8. Only in the medical field (as surgeon general) do officers reach grade O-9. These career fields are the most "stovepiped" in that horizontal occupational mobility is precluded and vertical mobility starts from a higher point and potentially ends at grade O-8 for most.[2]

Another broad category we have labeled technical and support. These positions include such fields as supply, ordnance, chemical, maintenance, transportation, engineering (other than combat), acquisition, and logistics. Officers in these fields typically enter as O-1s and move upward in them. However, officers from other career areas may enter these fields later in their career—for example, at grade O-3 or O-4. While it is possible to make the highest grade of O-10 within these fields, most officers have O-9 as the highest possi-

_____

[2]Horizontal mobility is movement to different positions at the same rank or grade level. Vertical mobility is movement up through the rank or grade hierarchy. The Secretary's concern is largely that horizontal mobility is too rapid, while the services have concerns about slowing vertical mobility as a means to decrease horizontal movement.

ble end point in their career. Horizontal and vertical mobility exists. For assignments, officers move horizontally and serve in positions in the Joint Staff, combatant commands, defense agencies, or the Office of the Secretary of Defense. Also, line officers sometimes serve in technical and support positions as a means of broadening their experience. Specialization provides a clear path for promotion and advancement, but the approximately 140 G/FOs developed in these communities do not get a balanced view of all aspects of the organization and thus may not be the best officers for general executive positions.[3] Seldom do officers in these communities rise to O-10.

In this report, we define the line communities as those associated with the direct application of and conduct of warfare operations.[4] These officers (approximately 660 G/FOs) lead and command at the tactical, operational, and strategic levels of war. They are students of military and naval science and create warfare strategy and doctrine at the service, joint, combined, and interagency levels. Ultimately, they will become combatant commanders, chiefs of service, or even the Chairman or Vice Chairman of the Joint Chiefs of Staff. These communities are broadly called combat arms (e.g., infantry, armor, artillery) in the Army and unrestricted line (e.g., surface, submarine, aviator) in the Navy. For the Air Force, the line includes pilots and navigators; in the Marine Corps, the line is unrestricted. Career patterns are similar for line officers across services. Officers enter at grade O-1, and their early career years are spent becoming experts at their particular disciplines, such as infantry, aviation, or surface warfare. This specialized expertise and cultural understanding can be the basis of a career through grade O-6 (and a few to grade O-7). However, officers must gain broader organizational skills and a

---

[3]Drucker (1954). We are indebted to fellow RAND researcher John Boon for identifying much of the relevant literature.

[4]The "line" is a well-established title with lengthy history or cultural acceptance in some services and is not used specifically in others. We use the term to classify a set of unique military skills generally acquired through established military education, training, and experience. Those in professional, technical, and support communities are not likely to make grade O-10 part of their career paths and are not considered to be part of the line communities.

deeper understanding of the service (not the community) culture to sustain vertical mobility. Learning the military service's organizational skills and embracing its culture can lead to a career pattern that includes promotion to grade O-9. Reaching the highest grade level (O-10) requires learning and adapting within a broader corporate set of skills and culture—"jointness" or the national security environment writ large. This level includes the service chiefs, the Chairman and Vice Chairman of the Joint Chiefs of Staff, the combatant commanders, and the specialized senior billets, such as the doctrine, training and education, nuclear, and the service component commands. Last, there are a few unique positions in which officers trade vertical mobility for long service in positions where their more narrow military experience is valued. These positions include service school and academy faculty and staff.

Horizontal mobility is largely unlimited within the four career patterns outlined for the line community. Officers can and do move between positions within their service and in the larger national security establishment. Vertical mobility caps near the points shown in Figure 2.1. The ability to go higher depends on one's success within the career patterns shown. Many officers succeed within the environment of their service and its culture but do not have the ability or opportunity to thrive in the larger national security environment. This report focuses on officers who serve in the line communities. As shown in the figure, these officers have broad horizontal mobility because they might serve in line, technical, or support positions. Vertical mobility is primarily within line positions. In the line community, career paths are not as well defined as they are in the other communities, and avenues for advancement may not be clear. Our empirical look at O-10 careers in Chapter Four confirms that there are many paths to the top. Officers in these communities learn to balance many functional considerations through job transfers among the organization's functions. Those who rise to the most-senior positions appear to have executive qualifications, the adaptability to develop and capitalize on these qualifications, and the opportunity to get a broad and balanced view of the organization.

SENIOR OFFICERS FLOW RAPIDLY THROUGH THE SYSTEM[5]

An initial data assessment of the promotion patterns and career tenures of G/FOs confirms that senior officers retire relatively early[6] and are able to do so by moving relatively rapidly through both assignments and ranks. For example, Figure 2.2 indicates that, before promotion, officers spend approximately three years as O-7s, two to two-and-a-half years as O-8s, and two-and-a-half years as O-9s. The three years at O-7 typically splits between two 18-month assignments in some services and either 12 months or two years in others; then most officers will fill one to two assignments at each subsequent pay grade. Officers promoted upward show slightly different assignment

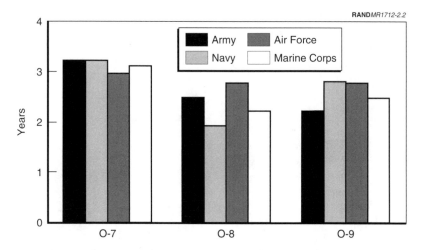

SOURCE: DIOR, General and Flag Officer Database. Mean years in grade for all G/FOs promoted to O-7 on or after January 1, 1987.

Figure 2.2—Average Number of Years Before Next Promotion

---

[5]The data portrayed in this section encompasses all military G/FOs. The focus on G/FOs in line communities begins in Chapter Four.

[6]O-10s serve, on average, 3.5 years as O-10s, and 87 percent of them retire before reaching age 60; in contrast, CEOs serve, on average, 8.4 years, and 69 percent of them stay past age 60 (Hadlock, Lee, and Parrino, 2002; Lucier, Spiegel, and Schuyt, 2002).

patterns: They also tend to serve two assignments at grade O-7 but are slightly more likely to have only one assignment at O-8 and O-9 on their way to O-10. Thus, the result is the assignment pattern shown in Figure 2.3, in which retiring O-10s typically show a total of five to six assignments. These assignments are more likely to be longer assignments (greater than 29 months) at the senior pay grades: Almost half of O-10 assignments are longer than 29 months, while only 16 percent of past O-7 assignments have exceeded the same amount of time (Figure 2.4).

G/FOs tend to retire from the military with approximately 30–35 years of service. Figure 2.5 indicates that O-7s tend to have closer to 29 or 30 years of service and O-10s leave with only about five more years of military service. This pattern is possible because those destined for eventual promotion to O-10 tend to get promoted to O-7 sooner than their peers do (Figure 2.6). The cumulative effects of this pattern appear in Figure 2.7, which shows that most retiring O-10s have spent approximately 10 years as a G/FO, while departing O-7s have spent an average of three years as a G/FO.

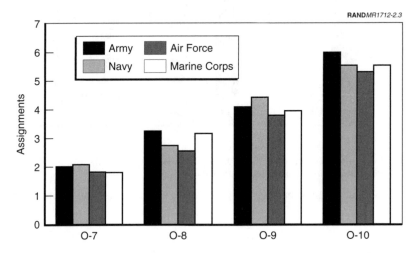

SOURCE: DIOR, General and Flag Officer Database. Mean number of assignments as a G/FO before retirement for all G/FOs who retired on or after January 1, 1980.

Figure 2.3—Average Number of Assignments at Retirement

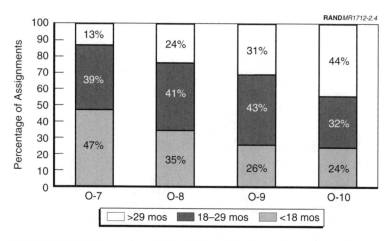

SOURCE: DIOR, General and Flag Officer Database. Percentage of G/FO assignments beginning on or after January 1, 1990. (Data may not sum to 100 percent due to rounding.)

Figure 2.4—Length of General and Flag Officer Assignments

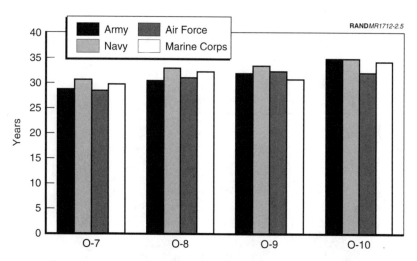

SOURCE: Date of retirement taken from DIOR's General and Flag Officer Database. Basic Active Service Date taken from JDAMIS. Mean number of years between date of retirement and Basic Active Service Date. Chart includes all officers who were (1) promoted to O-7 on or after January 1, 1980, and (2) included in the JDAMIS database.

Figure 2.5—Average Time in Service at Retirement

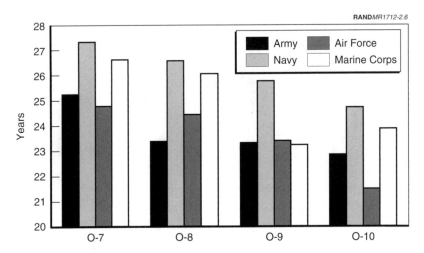

SOURCE: List of officers and date of promotion to O-7 taken from DIOR's General and Flag Officer Database. Basic Active Service Date taken from JDAMIS. Mean number of years between date of promotion to O-7 and Basic Active Service Date. Chart includes all officers who were (1) promoted to O-7 on or after January 1, 1980, and (2) included in the JDAMIS database.

Figure 2.6—Average Time in Service Before Promotion to O-7

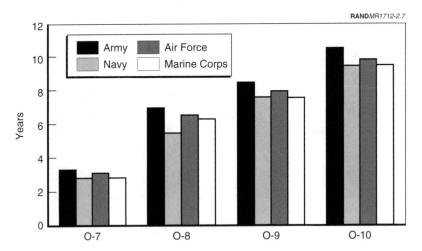

SOURCE: Taken from DIOR's General and Flag Officer Database. Mean number of years between retirement date and date of promotion to O-7 for all G/FOs who retired on or after January 1, 1980.

Figure 2.7—Average Total Years as a General or Flag Officer

SUMMARY

Consistent with expressed concerns, while assignments get slightly longer at the higher G/FO pay grades, most are shorter than 30 months. G/FOs promoted to the highest ranks tend to have had two assignments while at O-7 but only one at higher pay grades. This is consistent with quick promotion through the pay grades. While officers spend three years at O-7 (split between two assignments), they spend only two to two-and-a-half years at O-8 and two-and-a-half years as O-9 before promotion. At the conclusion of their career, retiring O-10s spend approximately 10 years as G/FOs and only three-and-a-half years as O-10s. The average O-10 retires with approximately 33–34 years of military commissioned service. Other retiring G/FOs depart with similar amounts of commissioned service because officers eventually promoted to O-10 typically are promoted to O-7 at an earlier point in their careers than are most new G/FOs.

Chapter Three

A FRAMEWORK TO ANALYZE AND MODEL THE GENERAL AND FLAG OFFICER MANAGEMENT SYSTEM

In this chapter, we present a generalized career framework that we will later employ to analyze and model career management practices for G/FOs. This framework combines prior theoretical work and empirical studies from the human resource literature. Some concepts in our framework have already been used by other authors to describe and analyze military careers; other concepts are taken from the corporate world but can justifiably be applied to military careers. Senior military officers often enter successful second careers in the private sector and tend to resemble corporate executives in terms of personality.[1] Corporate and military organizations themselves are similar in their use of periodic job rotations as a means to provide developmental opportunities for future executives and to test their skills and abilities. The concepts are also robust across a variety of corporate settings and for executives with different backgrounds, which suggests that they may be common even outside the private sector.

CUMULATIVE LEARNING THROUGH WORK EXPERIENCE

We base our career framework on the notion that work experience accumulates through a series of assignments that ideally prepare a person for increasingly demanding and complex jobs. An organiza-

---

[1]Personal communication with clinical psychologist Otto Kroeger. Kroeger is a distinguished visiting lecturer at the Army War College and has taught leadership development at the intermediate and senior service schools for years. Based on his extensive study of personality types, he says that senior military officers and corporate executives are indistinguishable.

tion, therefore, might define a career as "a sequence of work roles that are related to each other in a rational way so that some of the knowledge and experience acquired in one role is used in the next."[2] This definition assumes a logic to the job sequence so that it creates a skill set valuable to the organization.

The knowledge and experience gained through the typical job sequence follows a predictable pattern: Early assignments generally build specific functional skills, general and often tacit organizational knowledge, and idiosyncratic personal insights. Later assignments tend to have more complex and ambiguous responsibilities that require application of functional, organizational, and personal knowledge gained in earlier assignments.

Executives report in *The Lessons of Experience: How Successful Executives Develop on the Job*[3] that their early assignments as executives or before that tended to have core elements that were fairly simple, providing only a few basic managerial challenges. Their early assignments also provided a high degree of organizational and personal learning. As they moved to higher organizational levels, these executives found their technical and functional skills less helpful because of greater ambiguities in their responsibilities, because they were assigned to different functional units, and because they moved from line to staff positions, which were more conceptual and strategic than tactical. In these positions, executives learned to gather and synthesize information and make decisions that involved some guesswork. They also had exposure to corporate culture, attitudes toward risk, and the broader context within which decisions are made. The authors of the study concluded that such assignments effect a mental transition from thinking tactically to thinking strategically.

These identifiable patterns suggest that accumulated experience is not serendipitous. On the contrary, corporations actively manage the careers of their high-potential employees. A study of career management[4] examined 33 large U.S. corporations, with each reporting

---

[2]Morrison and Hock (1986, p. 237).

[3]McCall, Lombardo, and Morrison (1989).

[4]Derr, Jones, and Toomey (1989).

that it had some sort of program for identifying and developing future corporate leaders. Job rotation was by far the most common and most important developmental tool in these programs, and assignments typically lasted two to three years; assignments for other employees were longer, meaning job rotation was less common. This corroborates other research not specifically focused on high-potential employee programs that found job rotations are more common among employees who are performing well.[5]

Taking the basic concept of cumulative experience a step further, Morrison and Hock argue that, as a person moves from assignment to assignment, the magnitude and direction of influence between individual and assignment gradually change.[6] They draw their conclusion based on studies of various occupations—including naval surface warfare officers, who progress from division officer to department head to executive officer to commanding officer. Along the way, an officer's responsibilities become more complex and less prescribed; eventually, the Navy depends heavily on the officer's good judgment to command a ship. By that time, a commanding officer can draw upon the knowledge, experience, organizational understanding, and personal growth nurtured over his 20-year career. The pattern is similar to that of the corporate executives who found that later assignments had more ambiguous responsibilities and required more guesswork, making personal judgment and experience more important. To put it simply, early in a career the job shapes the person, but later in a career the person shapes the job.

CAREER STRUCTURE

Morrison and Hock posit a career development model in which a sequence of assignments prepares a person for a "target position."[7] We propose a taxonomy of "developing jobs" and "using jobs," in which the former type of job confers knowledge, skills, and personal growth that are put to use in the latter. Almost all jobs have some "using" aspect, in the sense that organizations want to put to use the

---

[5]Campion, Cheraskin, and Stevens (1994).

[6]Morrison and Hock (1986, p. 240).

[7]Morrison and Hock (1986, p. 241).

incumbent's existing knowledge, skills, and abilities. The key distinction, therefore, is the extent to which the job prepares a person for a higher-level assignment. By this definition, many using jobs logically occur at career's end, regardless of how high a person has advanced in an organization. For those who have risen to executive levels of management, their using jobs are critical to the success of the organization and demand a high degree of accountability and stability—the most obvious example in the private sector being the position of CEO. In positions such as these, executives shape not only the job itself, but often the entire organization.

Senior corporate and military leaders pass through multiple sequences of developing and using jobs. In *Corporate Mobility and Paths to the Top*, Forbes and Piercy identify a consistent development pattern in the careers of 230 chief executives of large industrial firms: Jobs or responsibilities changed rapidly early in a career, stabilized when the future CEOs reached middle management, and again changed rapidly when these individuals became junior executives.[8] As senior executives, they continued to have high-level developing jobs that either groomed them for promotion to CEO or at least put them in competition for that position.[9] In the military, the vast majority of officers who promote to colonel or Navy captain end their careers in those ranks, but a small number become G/FOs, setting an even smaller number on course to become combatant commanders, service chiefs, or even Chairman of the Joint Chiefs of Staff. Before reaching one of these positions, G/FOs go through another round of schooling, serve as commanding officer of a division or battle group or numbered air force, and may have special assignments to NATO, the White House, or the Pentagon. These are important positions that require high levels of knowledge, experience, and so forth, but are nonetheless part of a developmental process that teaches strategic thinking, prudent risk taking, knowledge of the international political environment, and other elements needed in senior military executives.

---

[8]Forbes and Piercy (1991).

[9]The executives studied in *The Lessons of Experience* reported that key developmental assignments at senior levels include starting a new plant or operation from scratch, turning around an existing operation in trouble, or making a leap in scope that entails managing more people, dollars, and functions (McCall et al., pp. 42–57).

LEARNING AND ACTION ON THE JOB

The developing job–using job distinction provides a framework for analyzing when learning and action take place within a career. We also consider when learning and action occur within an assignment.

Research has shown that executives follow predictable patterns in the timing of learning and decisionmaking in a new assignment. Based on longitudinal studies of high-level executives in large corporate units, Gabarro identified five stages of learning and action that executives move through after entering a new position:[10]

1. **Taking Hold**—a period of orientation and evaluative learning and corrective action that lasts three to six months.

2. **Immersion**—a period of relatively little change but of more reflective and penetrating learning; this stage could require nearly a year following the Taking Hold stage.

3. **Reshaping**—a period of major change during which the new manager acts on the deeper understanding gained in the preceding stage; lasts about six months.

4. **Consolidation**—a period in which earlier changes are consolidated; lasts four to eight months.

5. **Refinement**—a period of fine-tuning and relatively little major additional learning.

After entering a new managerial position, executives need between two-and-a-half and three years to progress through these stages, beyond which they are no longer considered new to their position and enjoy deep influence and knowledge. This pattern is robust across a variety of situations, including well-run organizations, with the differences being in magnitude of the learning and action, rather than in the timing or sequence of the stages. Furthermore, Gabarro found that reshaping, consolidation, and refinement are not limited to the responsibilities of the CEO; although lower-level executives and managers do not have as much authority over structural or

---

[10]Gabarro (1987).

strategic issues, they nonetheless have sufficient latitude to have a meaningful effect on the organizations they manage.

Combining the developing job–using job concept with the stages of learning and action, we conclude that not only the sequence but also the duration of assignments affects an executive's usefulness in a position. For executives to exert their influence in using jobs, they must remain in the assignments long enough to learn the new role, take meaningful action, and be held accountable for their actions. But organizations must balance effectiveness and accountability with other goals, such as executive development, promotion opportunity, promotion selectivity, and job satisfaction. If high developmental value is ascribed to certain jobs, organizations may choose to sacrifice some effectiveness and accountability, which could lead to shorter assignments. Recall that high-potential employees change jobs faster than their peers do, typically every two to three years until they reach the later stages of their careers. Gabarro's study suggests that this is sufficient time for people to learn a job but not enough to allow them to reach a state of deep influence and knowledge. However, corporations evidently make accountability and stability a high priority in critical using jobs, in which an executive could serve many years if his performance is good. CEOs average more than eight years on the job, and almost 70 percent serve to age 60 or older.[11]

APPLYING THE DEVELOPING JOB–USING JOB FRAMEWORK

In this chapter, we have offered a framework to examine careers—specifically, as a combination of developing jobs and using jobs. Few jobs within an organization are purely one or the other, but the terms serve as a useful distinction nevertheless because the characteristics of a job have implications for, among other things, when an assignment occurs in a person's career and how long the assignment

---

[11]Based on a study of CEO turnover between 1971 and 1994 (Lucier, Spiegel, and Schuyt, 2002). Tenure and age at retirement of CEOs in 15 industry groups were compared with those of CEOs in regulated industries using *Forbes* executive compensation surveys, Standard & Poor's COMPUSTAT database, and succession announcements reported in the *Wall Street Journal*. The data we report are for CEOs in the 15 industry groups; data for CEOs in regulated industries are not much different (Hadlock, Lee, and Parrino, 2002).

should be. The length and timing of assignments, in turn, affect a variety of competing organizational goals, such as effectiveness, accountability, development, opportunity, selectivity, and personnel job satisfaction.

The subsequent analysis and modeling of G/FO careers apply the developing job–using job career framework and demonstrate how these competing interests might be balanced. We assume that freshly minted O-7s are entering a new developmental phase, albeit at a relatively high level, that will prepare some of them to become O-10s, the most-senior leaders in the military. As such, they can have tremendous influence on their respective service and on national military strategy by serving in the ultimate using jobs.

CURRENT GENERAL AND FLAG OFFICER DEVELOPMENT IN THE DEVELOPING JOB– USING JOB FRAMEWORK

Organizational practices in the military and in the private sector indicate that the developing job–using job framework provides an effective career model. However, developing jobs and using jobs probably cannot be defined in any manner other than very general terms because they depend on the type of organization and the skills valued in its executives. Furthermore, most jobs do not fit neatly into one category or the other. Regional vice presidents or division commanders, for example, are in positions important to their organizations but also in those that lead to CEO or Chief of Staff of the Army. Ultimately, developing jobs and using jobs are whatever an organization wants them to be. As outside observers, the best indicator we have of the nature of a job is its timing: We can be reasonably certain that CEOs and O-10s are in using jobs and that junior managers and O-1s are in developing jobs; however, it is tougher to discern those in the middle.

These issues arose as we attempted to identify developing jobs and using jobs for G/FOs. It is far beyond the scope of this project for us to make value judgments about the desirability of specific skills and experiences in senior military leaders or to discern which jobs best confer them. However, we did want to get a picture of how the services develop and use G/FOs; therefore, we applied a few common-sense rules to categorize positions. Once we identified developing jobs and using jobs, we looked at how they compare across the services and at how they are managed in light of what we know of the corporate world. We present our findings in this chapter.

AN EMPIRICAL METHOD FOR IDENTIFYING DEVELOPING JOBS AND USING JOBS

Our initial analysis, as portrayed in the prior chapters, addressed all G/FOs. However, our sponsor asked us to focus on positions for line officers. For our analysis, we set aside officers who have careers within the non-line communities, such as professional, technical, and support, as described in Chapter Two. Based on the empirical data, we included in our analysis armor, infantry, and field artillery officers in the Army; unrestricted line officers in the Navy; Air Force pilots and navigators; and line officers in the Marine Corps.[1] These specialties have historically led to O-10. Once we had this subset of officers, we included in our analysis all assignments that officers in these specialties had as G/FOs, which included some assignments in technical, support, and, in a few instances, even professional positions.

We examined the assignment history of O-10s from these selected line communities to see how they were developed after promotion to O-7 (the assumption being that they had entered a second, high-level developmental stage upon becoming G/FOs). We applied our commonsense rules to this information. The underlying logic is that the timing and frequency of jobs provide a reasonable indication of whether they are developing or using. Thus, we assumed that all O-7 jobs are developing jobs and all O-10 jobs are using jobs. As for the jobs in the middle—those at O-8 and O-9—we assumed that the O-8 jobs that appeared frequently on O-9 or O-10 resumes and O-9 jobs that appeared frequently on O-10 resumes are developing jobs. These rules were designed to be conservative in identifying using jobs because all G/FOs are eligible for promotion or a new assignment, so in theory anything short of Chairman of the Joint Chiefs of Staff could be considered a developing job.

---

[1] We recognize that "general" officers are no longer considered to be part of a particular branch, so this is really a description of their background as company- and field-grade officers. There were approximately 2,500 officers and 1,150 positions (many no longer in existence) included in our analysis, reflecting more than two decades of data. Approximately 540 officers are in these specialties in any given year. All further references to the line are to this more narrowly defined group of officers.

Before providing the details of how we separated O-8 and O-9 positions into developing jobs and using jobs, we note an important caveat: This analysis is descriptive, not prescriptive. We believe it is useful to identify developing jobs and using jobs at these grades because it sheds more light on how the careers of senior military leaders are managed and because we use that information in our modeling efforts, as discussed in later chapters. But the services should identify developing jobs and using jobs following a different process. In short, we are neither recommending which jobs should be identified as developing jobs nor recommending a process by which the services should identify these jobs.

For each position, we know the officer's unit and job title (when dual-hatted, officers have multiple units, multiple job titles, or both). Using this information, we grouped positions into 11 broad job categories based on the type of unit an officer was assigned to and the

Table 4.1

Categories of General and Flag Officer Positions

Organization Type	Position Type
Operations	Command
	Deputy Command
	Staff
Training and Education Center	All
International or Multi-organizational	Liaison
Joint	Command
	Staff
Headquarters	Command
	Staff
Professional	Command
	Staff

NOTE: Some of the categories—e.g., Professional Command and Professional Staff—do not generally apply to line officers. These categories were included as a framework appropriate to all G/FOs. Still, there were some positions that fell into these categories.

duties associated with the job title, as broken out in Table 4.1. By design, the categories are not service-specific, which facilitates comparisons but may limit their usefulness to an individual service conducting its own analysis of officer positions. With only 11 categories and 1,811 unique combinations of units and job titles in our database, most of the groupings are rather large, so it would have been possible to sort positions into additional categories. However, as the number of categories increases, the distinctions between them become less clear, and even with only 11 categories there is occasional ambiguity in how a job is classified.

Next, we determined how often each job category appears on officers' resumes. If less than 10 percent of O-10s had a particular category of job while they were O-9s, we considered that category "low frequency" for grade O-9. Likewise, if less than 10 percent of O-9s and O-10s had a particular category of job while they were O-8s, we considered that category "low frequency" for grade O-8. The 10-percent cutoff could have been set higher or lower, which, of course, would have resulted in either fewer or more categories being counted as "low frequency."[2] All categories that surpassed the 10-percent threshold are considered "high frequency" for grades O-8 or O-9.[3]

All jobs that fell into the low-frequency categories (appearing in fewer than 10 percent of officers' resumes) we counted as using jobs and set them aside. We returned to the jobs in the high-frequency categories and identified those that rarely, if ever, are filled by an officer who will subsequently be promoted. These, too, are using jobs because they typically appear at the end of an officer's career. Again, we established some rules for identifying career end points. Specifically, if no officer who has held a particular job since 1990 (including officers assigned in the late 1980s and still there in January 1990) has subsequently been promoted, that job is considered a career end point and therefore a using job.

---

[2]We recognize that 10 percent may not seem to be a particularly high cutoff, but there are 11 job categories, and officers only have one or two assignments apiece at grades O-8 and O-9. If the categories were of equal size, we would expect only to see each one on about 10–20 percent of officer resumes.

[3]"High-frequency" is merely a convenient nomenclature for "not low-frequency."

To summarize our methodology, it can be thought of as a pair of filters that separate using jobs from developing jobs at grades O-8 and O-9. The first filter is applied to job categories, sorting them into high- and low-frequency categories. The second filter is applied to individual jobs within the high-frequency categories to identify career end points. Jobs that are in low-frequency categories or that are career end points we count as using jobs; those that are in the high-frequency categories and are not career end points we count as developing jobs. We applied these filters separately for O-8 and O-9 jobs and for each service. Figure 4.1 illustrates the methodology. Upon obtaining our lists of developing jobs and using jobs, we counted the number of such jobs filled each month since January 1990 to determine the percentage of developing jobs and using jobs for each of the services.[4] The results of this analysis are shown in Chapter Five.

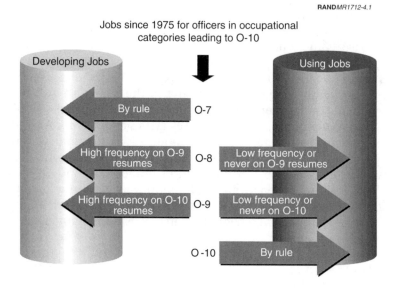

RAND*MR1712-4.1*

Jobs since 1975 for officers in occupational categories leading to O-10

Developing Jobs

Using Jobs

By rule — O-7

High frequency on O-9 resumes — O-8 — Low frequency or never on O-9 resumes

High frequency on O-10 resumes — O-9 — Low frequency or never on O-10

O-10 — By rule

Figure 4.1—Summary of Process to Determine Developing Jobs
and Using Jobs

[4]Our grouping of developing and using positions is available to service personnel by inquiry to the authors. Contact information is located in the preface of this document.

THE LENGTH OF DEVELOPING JOB AND USING JOB ASSIGNMENTS

In the previous chapter, we examined the linkages between job type (developing or using) and assignment tenure. Executives require several years in an assignment to reach a point of effectiveness, accountability, and deep influence and knowledge. Long job tenures are typical of CEOs and other senior executives who occupy critical using jobs. Earlier in their careers, however, these executives likely moved into a new developing job every two or three years as they were being groomed as future leaders. Applying Gabarro's finding that managers and executives go through predictable stages of learning and action when they enter a new position, we conclude that two years might be a sufficient amount of time for a person to acquire new skills and knowledge in a position but not enough for significant action and accountability.

Now that we have identified developing jobs and using jobs for G/FOs, the natural question is whether the services currently link job type to assignment tenure. Figure 4.2 presents the median duration, in months, of developing job and using job assignments at grades O-8 and O-9 since 1990. In each of the services, O-9 assignment tenure is slightly longer than O-8, and in the Army, Navy, and Air Force, using assignment tenure is slightly longer than developing assignment tenure, but the differences are small. Only for Air Force O-9s are using assignments even six months longer than developing assignments.[5] Thus, we conclude that there is little connection between job type and job duration at grades O-8 and O-9. The median assignment length in an O-10 job ranges from a low of 26 months for the Air Force to a high of 35 months for the Marine Corps, which is generally not much longer than O-9 using assignments.[6]

---

[5]Data for the Marines, particularly those for using jobs, are based on very small numbers. Not only is the Marine Corps itself much smaller than the Army, Navy, or Air Force, but its percentage of using jobs is also smaller. The median duration of O-9 using jobs, for example, is based on only five observations.

[6]We report averages for most data and modeling results, but in this case we use the median because assignment lengths tend to have outliers, particularly on the low end of the distribution. Out of 256 Air Force developing assignments at O-8, for example, 17 lasted less than six months and 54 lasted less than a year. The median is less sensitive than the mean to outliers and, we believe in this case, provides a more accurate description of the central tendency of assignment lengths. Ultimately, the differences

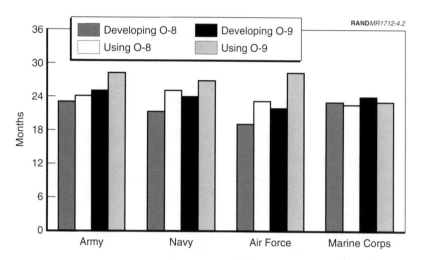

SOURCE: DIOR, General and Flag Officer Database.

**Figure 4.2—Median Assignment Length in Developing Jobs and
Using Jobs Since 1990**

Another basis of comparison for O-10s is their private-sector counterparts, CEOs. Research based on *Forbes* executive compensation
surveys from 1971 to 1995 found that CEOs average more than eight
years on the job and that almost 70 percent serve to age 60 or older
(see Table 4.2).[7] By contrast, O-10s average only three-and-a-half
years in grade (which could include more than one assignment), and
nearly 90 percent retire before reaching age 60. Also of note is the
reason for CEO departures. Fifteen percent of CEOs leave voluntarily;
15 percent depart as the result of organizational mergers; and 15 percent of departures are performance-related. The majority of CEOs
(55 percent) retire from their position in compliance with age-related
corporate policy. Most O-10s, however, never reach statutory retirement age.

---

are small. For average assignment lengths, we make the same observation that only in
one case are using jobs at least six months longer than developing jobs at a particular
grade within one of the services.

[7] Hadlock, Lee, and Parrino (2002).

Table 4.2

Senior Executives' Job Tenure and Retirement Age (in years)

	O-10s	CEOs
Mean age when appointed	54.1	53.5
Mean tenure	3.5	8.4
Length of tenure (25th–75th percentile)	2.3–4.1	5–12
Those departing under age 60	87%	31%

SOURCE: CEO data from Hadlock, Lee, and Parrino (2002). O-10 data derived by RAND from the DIOR database, combined with the DMDC and JDAMIS databases.

THE DEVELOPING–USING FRAMEWORK IS THE BASIS FOR MODELING POLICY ALTERNATIVES

We introduced the developing–using job career framework and linked job type (developing or using) to assignment tenure. We used this framework to analyze G/FO careers. Each of the services has certain categories of jobs that we consider to be "low frequency" and therefore not typical of job rotations meant to develop officers for senior military positions. The services also have jobs that historically have been career end points, from which nobody who has served since at least 1990 has been subsequently promoted. Therefore, we conclude that it is reasonable to describe these jobs, as well as O-10 jobs filled by the most-experienced officers, as using jobs. Corporate practice and empirical research suggest that such jobs should be longer than developing jobs to make better use of officers' experience, to increase their accountability, and to improve the organization; however, we find that using assignments are scarcely longer than developing assignments.

These observations and findings are important inputs to our modeling efforts and form the basis for a variety of alternatives that we explore. The models, in turn, suggest how G/FO management policies could be revised and how promotion rates, promotion selectivity, time in grade, and career lengths would be affected.

Chapter Five

A REVISED MANAGEMENT SYSTEM: EFFECT ON GENERAL AND FLAG OFFICER DEVELOPMENT AND PROMOTION

The empirical analysis described in earlier chapters indicated that the premises of civilian career theory and executive development applied to G/FO development: Military leaders are assigned to both developing and using positions, although the tenure for these assignments does not vary. Thus, this analysis sought to ascertain whether the strengths of civilian executive development could be applied to future G/FO career paths by managing the tenure of developing assignments differently than using assignments. The intent was to stabilize some individuals and organizations by getting the maximum use from the most-senior G/FOs as well as those individuals not likely to be promoted further. The concern frequently cited by those who manage senior officers, as well as some in Congress, was that such stabilization would clog or cause stagnation in the system. The policy excursions described below increase stability in the system without reducing promotion opportunities for most other officers.

The basis of the policy excursions examined in this analysis was a variation in tenure between developing assignments and using assignments; the developing assignments were set to shorter durations than were the using assignments. Two-year-long developing assignments are similar in length to many current G/FO assignments. Maintaining two-year developing assignments will continue to allow sufficient numbers of officers to gain developing experiences and maximize the number of officers who are observed in such experiences.[1] These assignments emphasize the development of officers for subsequent, higher-level responsibility and judge their potential for such positions. Shorter developing assignments reduce organizational stability and reduce the likelihood of individual accountability but maintain selectivity for promotion. In contrast, longer using assignments maximize stability and accountability in certain organizations, without compromising developing opportunities for officers. Such assignments would also clarify an officer's expectations and permit him the opportunity to have a more significant effect on an organization at the height of his career.

MODELS SUPPORTED THE ANALYSIS

This analysis was conducted using two independent models and was subsequently validated on a third model. The primary model is a steady-state system dynamics model designed on the basis of stocks and flows, wherein the officers "flow" through stocks of developing and using assignments. The key model inputs include the mix of developing job and using assignments at each pay grade, the length of each kind of assignment, and the number of each kind of assignment officers receive at each grade. The outputs, which are calculated separately for officers in developing assignments and using assignments, include

- the number promoted (throughput)
- the promotion probability to each pay grade
- the probability of an O-7 reaching O-10
- the number of officers developed who are not promoted
- the average time in service overall
- the average time in each pay grade for those promoted

---

[1]This permits service flexibility and selectivity of officers for promotion. Since developing experiences entail a sacrifice of performance, the astute organization will try to minimize developing experiences, subject to the constraint that they provide just enough developing experiences to meet their long-term needs.

- the average time in each pay grade for those retiring

- the average time in job.

A second, more detailed model validated the findings of the primary model. We designed this entity-based model to permit an understanding of the particular assignments an individual receives and of the effects on others of fast-tracking certain individuals through the system. We also developed this model within the context of the research effort, but with different software and modeling techniques. A third, independently developed model also validated the model findings.

ANALYTICAL STEPS TO DETERMINE THE OPTIMUM CAREER MODEL

This analysis tested various G/FO management systems, based on the lessons learned from civilian executive management and the developing job and using job distinction. The analysis included the following steps and was conducted separately for each service:

1. Identify developing and using positions, by grade.

2. Set goals for time in position (e.g., two years in developing assignments).

3. Set goals for number and timing of positions (e.g., two using assignments for each O-10).

4. Assess feasibility of the system.

5. Assess "cost" of changes (i.e., any change in promotion probability).

6. Revise, from step 2, as necessary.

Identifying Developing and Using Positions, by Grade

For the purposes of this analysis, we used the developing and using distinctions that emerged from the analysis described earlier. Figure 5.1 shows the percentage of developing and using jobs, by service

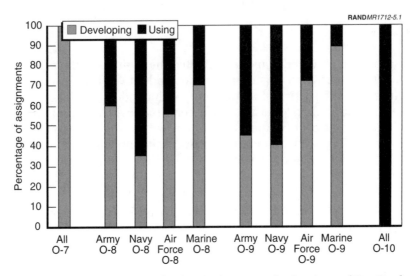

Figure 5.1—Developing and Using Assignments, by Service and Pay Grade

and pay grade. These percentages emerged empirically from the analysis but are not intended to be prescriptive. Instead, the services should review these allocations as part of policy implementation, as stated in Chapter Four, for several reasons. First, although we could observe how officers are currently being developed, it is not clear that these assignment patterns are the best way to develop officers for the future. Second, we cannot attribute causality to these patterns: Do officers who have a certain experience have a greater chance of promotion, or do officers with a greater chance of subsequent promotion receive certain assignments? Finally, the services should review our grouping of similar positions to identify any category they would treat differently. Nonetheless, the differences between the services' developing–using distinctions provide some inherent sensitivity analysis. Should the Army, for example, determine that it has many fewer developing positions for the future, its prospective system will likely look more like the Navy results described herein, given the limited number of developing positions empirically observed for the Navy.

MODELING AND ANALYSIS SUGGEST A NEW CAREER MODEL

The iterative modeling process described above evaluated 13 different cases for each of the services.[2] Some cases were judged infeasible. For example, if all O-7s serve two complete developing assignments (for a total of four years), the result would be insufficient numbers of promotions to O-8. In this instance, either O-8 positions go unfilled or officers are promoted to O-8 before completing two developing assignments.

The career model that emerged from the multiple excursions as best at balancing all concerns is one in which using assignments are four years long and developing assignments are two years long. O-7s and O-8s on the development track can serve in one or two developing jobs.[3] Officers promoted to O-9 will have served in a total of three developing jobs during their O-7 and O-8 tenures. Officers may retire or be separated after any of these developing jobs. O-9s likely to be promoted to O-10 will serve in a single developing job, and those not promoted to O-10 will retire at the conclusion of the developing job. Officers who serve in using positions at O-8 will serve in a single four-year assignment before retirement from O-8. Officers likely to retire from pay grade O-9 will serve in two using jobs at that grade. All O-10s will serve in two using jobs. The subsequent figures show the results of this analysis. (Appendix C includes the tabular format of these results, and Appendix D contains the graphical representations of the flows of officers through the system.)

This career model emerged as best because it satisfied the established outcome criteria of maximizing stability and accountability in some assignments without sacrificing considerable promotion opportunity.[4]

---

[2]The complete list of modeling excursions is included in Appendix B.

[3]That is, some officers will serve in one developing job as an O-7 and then complete two such assignments as an O-8. Other officers will serve in two developing job as an O-7 and then one as an O-8.

[4]The research sponsor office participated in judgment of "best" as meeting the established criteria.

MODELED OUTCOMES OF THE NEW CAREER MODEL

Promotion Throughput

Despite the concern from the services and some in Congress that slowing the system to increase accountability and stability in organizations would substantially reduce promotion opportunities, Figures 5.2–5.5 indicate that in most cases the annual number of officers promoted to O-7 increases compared with the status quo.[5] This is a result of some officers spending less time at grade O-7. However, O-7 assignments are longer, so even with less individual time in grade, organizations benefit from greater stability in O-7 positions. The number of officers promoted to grade O-8 is also approximately equal to (in the Army and the Marine Corps) or slightly greater than (in the Navy and the Air Force) the status quo. The number of officers promoted to O-9 increases for the Navy and the Air Force, remains roughly the same for the Marine Corps, and decreases (by one) for the Army. Promotions to O-10 decrease for all services.

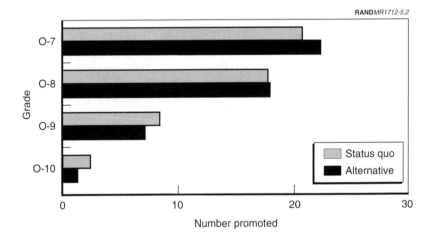

Figure 5.2—Army Promotions: Status Quo Compared with Alternative

---

[5]The number promoted is for the line community in each service as indicated, which is less than the total number promoted with which most readers are familiar.

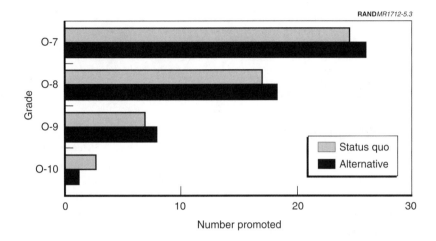

Figure 5.3—Navy Promotions: Status Quo Compared with Alternative

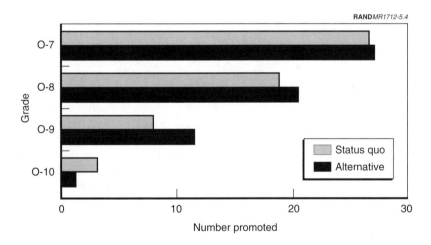

Figure 5.4—Air Force Promotions: Status Quo Compared with Alternative

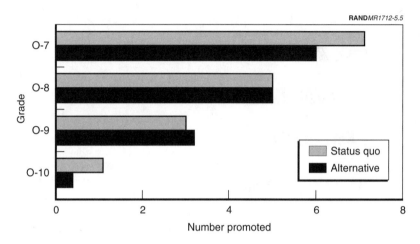

NOTE: The reader should note the different x-axis in the case of these results.

Figure 5.5—Marine Corps Promotions: Status Quo Compared with Alternative

Promotion Probability

Figures 5.6–5.9 indicate the likelihood of promotion for officers in the modeled alternative compared with current practice.[6] These figures note the likelihood that, for example, any officer promoted to O-7 will then be promoted to O-8. This promotion probability is indicated for both the overall population of officers in the alternative and also for those officers serving in developing assignments, as compared with the status quo. The probability of promotion to O-8 is slightly lower for Army officers compared with the status quo and slightly higher for Navy, Air Force, and Marine Corps officers. Promotion to O-8 is the same for developers and the total population because all O-7 assignments are considered developing jobs. Promotion probability to O-9 decreases slightly for the Army in total but increases for the total population in the other services, again compared with the sta-

---

[6]Promotion probability is defined as the number promoted from a grade divided by the number promoted to the grade. We calculated the status quo and alternatives on that basis. This is different from promotion opportunity, which uses the number in a promotion zone as the denominator.

tus quo. Developers are promoted to O-9 at a rate higher than the total population in the alternative and status quo. The likelihood of promotion to O-10 is less than that seen in current practice for both the total population and the developers.

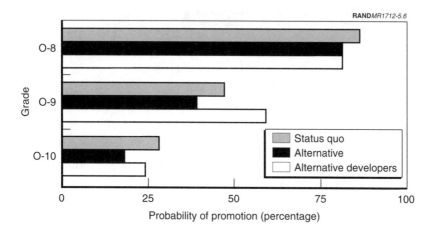

Figure 5.6—Army Promotion Probability: Status Quo Compared
with Alternative

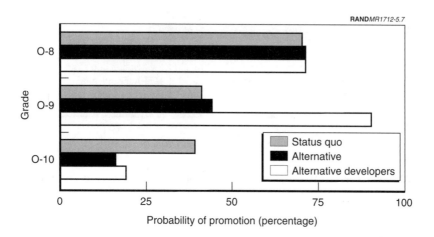

Figure 5.7—Navy Promotion Probability: Status Quo Compared
with Alternative

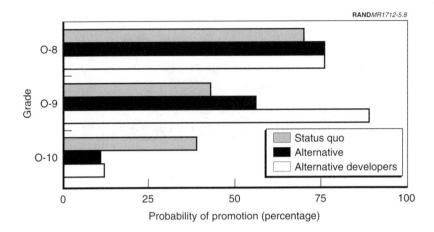

**Figure 5.8—Air Force Promotion Probability: Status Quo Compared
with Alternative**

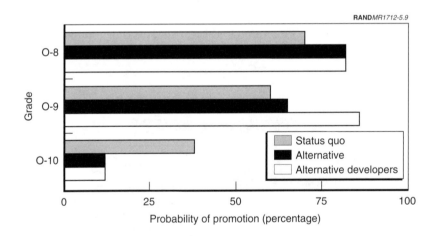

**Figure 5.9—Marine Corps Promotion Probability: Status Quo Compared
with Alternative**

Time in Grade at Retirement

Figures 5.10–5.13 display the average time in grade of officers retiring from that grade. For example, under the proposed career model, officers retiring from grade O-7 average approximately two-and-a-half years in grade O-7. This results from the retirement of officers who have one two-year assignment and the retirement of officers who have two two-year assignments. While this is less than the average O-7 time in grade for current retirees, the alternative is based on assignments longer than those of the current system, in which officers typically fill two 18-month assignments.

Average O-8 time in grade at retirement in the alternative is slightly longer for the Navy and the Air Force and shorter for the Army and the Marine Corps. However, the data are for officers serving in developing assignments[7] in the modeled alternative as well as those serving in four-year using assignments. Thus, the average time in grade for the modeled alternative tends not to reflect accurately the bimodal system of some O-8s and O-9s who serve two years in grade and others who serve four years (in the case of O-8s) or eight years (in the case of O-9s). Because all O-10 jobs are using jobs, all modeled officers promoted to O-10 will serve in two using assignments, for a total of eight years time in grade, which is considerably longer than the past average of three years time in grade for O-10s.[8]

---

[7]Officers not promoted to O-9 or to O-10 after serving in a two-year developing assignment would retire with two years in grade.

[8]The status quo data reflect officers who have retired; therefore, the relatively recent practice of assigning O-10s to a second O-10 assignment, as in the case of the Commandant of the Marine Corps becoming a combatant commander or of renominating a chief of service for additional tenure as with the Chief of Naval Operations, is not reflected in the data.

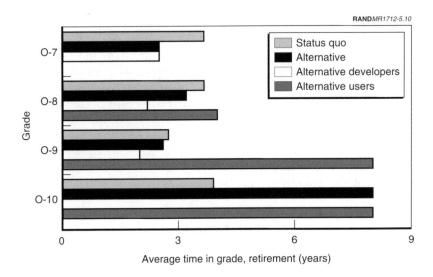

**Figure 5.10—Army Time in Grade at Retirement: Status Quo Compared
with Alternative**

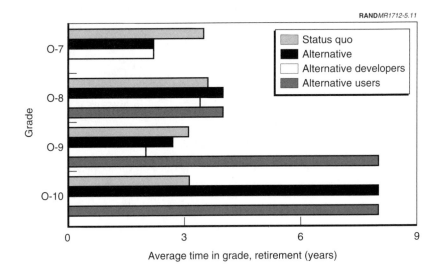

**Figure 5.11—Navy Time in Grade at Retirement: Status Quo Compared
with Alternative**

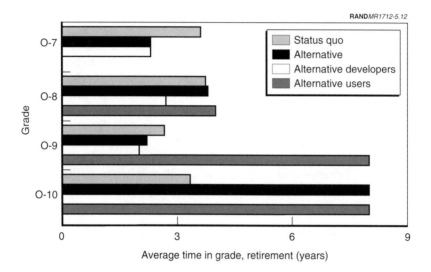

Figure 5.12—Air Force Time in Grade at Retirement: Status Quo Compared
with Alternative

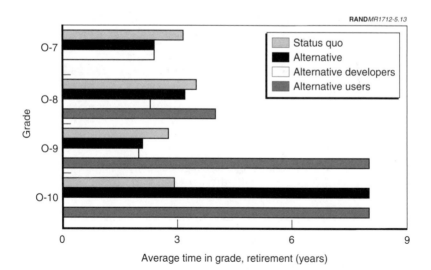

Figure 5.13—Marine Corps Time in Grade at Retirement: Status Quo
Compared with Alternative

Average Career Length at Retirement

Figures 5.14–5.17 show the average career length at retirement. The career length is based on our modeled time as a G/FO in addition to the average time at which officers are promoted to O-7.[9] While modeled O-7s who are not promoted to O-8 serve less time in the service than do currently retiring O-7s, officers at the other grades typically have similar (in the case of developers) or longer (in the case of users) military careers in the modeled alternative than in current practice. The longer careers are especially notable among O-9s in using assignments and O-10s, who are all serving in using assignments.

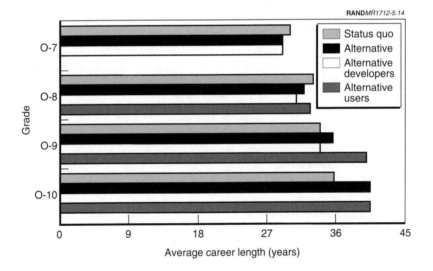

Figure 5.14—Army Career Length at Retirement: Status Quo Compared with Alternative

---

[9]In general, most services (with the exception of the Marine Corps) promote to O-7 soonest those officers who will eventually be promoted to O-10. This is discussed in more detail for line officers in Appendix A.

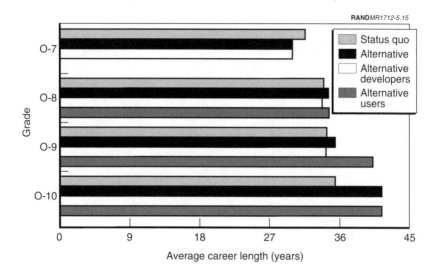

Figure 5.15—Navy Career Length at Retirement: Status Quo Compared
with Alternative

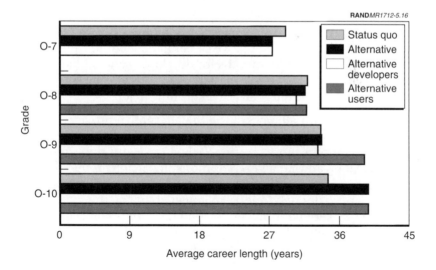

Figure 5.16—Air Force Career Length at Retirement: Status Quo Compared
with Alternative

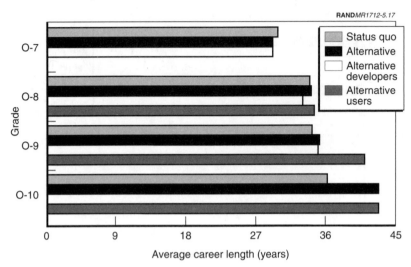

RAND*MR1712-5.17*

Figure 5.17—Marine Corps Career Length at Retirement: Status Quo
Compared with Alternative

Average Time in Job

Figures 5.18–5.21 compare average time in job for G/FOs in the status quo with that in the alternative career model. For the alternative, average time is always 24 months for developers and 48 months for users. The mix of developing jobs and using jobs at each service and grade determines the alternative average time in job.

For all services and all pay grades, the alternative provides greater time in job than the status quo. For Army O-7s, the alternative provides seven months more time in job than the status quo, compared with about four-and-a-half months more time in job for the Navy, Air Force, and Marine Corps. At O-8, the time in job for the alternative averages approximately one year longer than for the status quo. Because O-9s serving in using assignments in the alternative stay much longer than in the status quo, the average time in job at O-9 is longer in the alternative than in the status quo, even though some developers (in the Army and the Navy) will serve slightly less time in job than they do in the status quo.

By far, the largest increase in assignment tenure occurs at O-10: The alternative provides the Marine Corps with about 15 more months in

O-10 assignments compared with the status quo, the Navy with 19 more months, the Army with 20 more months, and the Air Force with 23 more months.

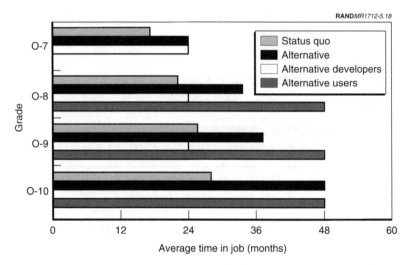

Figure 5.18—Army Average Time in Job: Status Quo Compared with Alternative

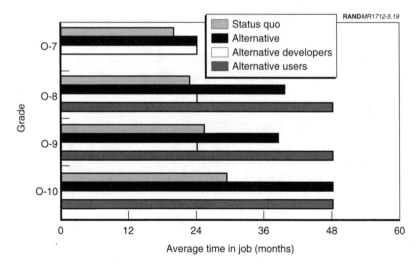

Figure 5.19—Navy Average Time in Job: Status Quo Compared with Alternative

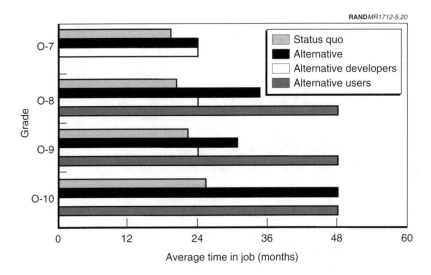

Figure 5.20—Air Force Average Time in Job: Status Quo Compared
with Alternative

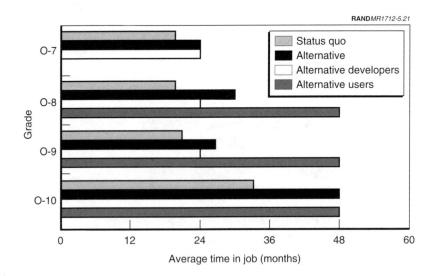

Figure 5.21—Marine Corps Average Time in Job: Status Quo Compared
with Alternative

SUMMARY OF MODELED OUTCOMES OF THE NEW CAREER MODEL[10]

Most of the services will experience a greater number of officers promoted to O-7 annually. (The Marine Corps will see one fewer.) Equal or greater numbers of officers will also reach grades O-8 and O-9, except one fewer to Army O-9. About half as many will rise to O-10, as the length of time that O-10s serve before retirement nearly doubles. Average career length will increase for all pay grades except O-7; however, O-7s will spend longer time in assignments than they do in the status quo. Average time in job will increase for all pay grades. Organizations will benefit from the stability of leadership tenures equal or longer than those witnessed today. And individuals will have clearer expectations about their future and, at the apex of their careers, an opportunity to produce a more significant organizational impact.

---

[10]While not an explicit consideration in our analysis, we assessed whether the new career model, including changed retirement compensation, would be more costly or less costly than the status quo. We examined the difference in total life-cycle costs in a steady-state, present-value framework. First, we estimated for each service the cost of an officer by grade and years of service. We then found—for both historical data and the model runs—the distribution of officers by grade and years of service. Comparisons reveal little difference between historical data and the new system. In the new system, the lower costs of fewer officers retiring at O-6, O-9, and O-10 are almost balanced with additional costs of retirees at O-7 and O-8. When the costs of positions being filled by more experienced officers are included, the net is near zero. Given that the grade structure does not change, the only relevant changes from a cost perspective are how experienced officers are when filling a job and how many officers are needed to fill all jobs. When a job is filled with a more experienced officer, the cost of filling that job increases. However, when officers are permitted to have more assignments in a grade and thus stay longer at that grade, fewer officers need to be promoted to that grade, thus decreasing the number who retire from that grade and decreasing retirement costs.

Chapter Six

REACTIONS TO THE CURRENT SYSTEM AND PROPOSED SYSTEM

Our research and analysis included interviews with serving and retired G/FOs to determine their perceptions of the current management system, as well as to predict likely responses to a changed system. In the course of this analysis, we also spoke with experts who manage the current system and accumulated their concerns about possible changes. This chapter describes the interview process and protocol and then addresses concerns through a combination of interview responses and modeling and analysis. The chapter concludes with additional salient observations that emerged from interviews conducted with senior officers.

INTERVIEWS WITH GENERAL AND FLAG OFFICERS

We conducted approximately 20 interviews with serving and retired G/FOs, primarily O-9s and O-10s, and with senior civilians. The interviews ranged from 30 to 60 minutes and were guided by a semi-structured format. In other words, we explained the interview the same way to each participant, but the content and the direction of the discussion were flexible, based on the interests, attitudes, opinions, availability, and enthusiasm of the participants. We sought personal views on G/FO management, based on the interviewee's experience and personal perspective. The questions we asked were broad and meant to elicit views about the matters at hand; they covered G/FO careers, assignments, management goals, development, and retention. We also asked interviewees not to feel restricted if they had observations about career management not addressed by the questions. We do not claim that the interviews were representative but do believe that they provide excellent insights into the current and prospective management of G/FOs.

ADDRESSING CONCERNS ABOUT MANAGEMENT CHANGES

During the course of our research, several concerns were raised about repercussions from the proposed management change. We address these below, frequently using observations from our interviews.

Retention

One concern is that officers would not stay for longer careers or would not stay longer in assignments perceived to be career end points. Our interviews, however, suggest that this is not a valid concern, and we conclude from them that retention is not likely to be a problem. There will always be voluntary leavers and unexpected retirements. The retention issue is probably more fit to each individual officer rather than the system as a whole.[1] The drivers for leaving early tend to be either family issues or outside employment offers too attractive to refuse. However, some officers do become too tired and should leave before performance becomes a problem.

Moreover, the system needs fewer officers to commit to staying longer, so not everyone need agree to longer tenure. Further, should officers not stay as long as desired, the average assignment length and career tenure might not be as long as desired, but the system will still be manageable. For example, in the Army alternative, seven officers are promoted each year to O-9, and roughly 11 officers retire as O-8s who were not promoted to O-9. Should more O-9s retire without serving complete assignments, the promotion probability to O-9 would increase, but there are sufficient numbers of officers who

---

[1] We observed a high level of organizational commitment as we discussed retention. Organizational commitment is " . . . the relative strength of an individual's identification with and involvement in a particular organization. Conceptually, it can be characterized by at least three factors: a) a strong belief in and acceptance of the organization's goals and values; b) a willingness to exert considerable effort on behalf of the organization; and c) a strong desire to maintain membership in the organization" (Mowday, Steers, and Porter, 1979).

could be promoted to O-9 rather than retiring as O-8s. This is even more the case for promotion to O-10. Should O-10s serve half as long as the alternative suggests, the number promoted annually to O-10 could double, and that promotion throughput would resemble today's numbers.

A related issue is whether all O-10s have the appropriate portfolio of experiences for a second O-10 assignment. If, for this reason, only half of those selected for O-10 receive a second four-year assignment, the result would be the same as that discussed above. The number promoted to O-10 would increase, but there are sufficient departing O-9s to accommodate this increase, and the number of officers promoted to O-10 would be about 75 percent of today's numbers.

We anticipate, based on the interviews, that officers will stay for longer assignments and longer careers. Should they behave differently than predicted, the system may not achieve all the increases in stability and accountability—but it would look no worse than today's system.

Flexibility

We agree with concerns raised in the interviews that the system must remain flexible. Thus, we assert that an improved system should not be overly rule bound: Performance and logic matter. Short or long length of career and length of assignment are symptoms. They are means to some end, but what is the end? Interviewees suggested such intentions as establishing control, achieving better organizational performance, and creating military leadership talent. It may be useful to be as explicit as possible about the ends. Length should not be the only metric; G/FO management needs to be about performance, not just calendars.[2] G/FOs should not be retained just to get maximum service; maximum contribution is also important. Moreover, some positions cram more experience into shorter periods—for example, a corps commander who is also a joint task force commander. Time is important to changing organizational cultures, to innovation, and to introducing change and seeing results. Organizations

---

[2]From an analytical perspective, peaks and valleys in retirement patterns could cause promotion probability spikes and assignment timing problems in an inflexible system.

need more individual accountability, and G/FOs want time to see their efforts bear fruit in organizations. Senior officers spoke regretfully of their short assignment tenures. One problem that is mitigated by our recommended changes is the inflexibility of the current system, which is structured on a routinization of officer movement every two years.

Compensation

Another valid concern is that the current compensation structure is inadequate for longer G/FO careers, which we also address in our concluding chapter and Appendix E. Many of the senior officers we interviewed mentioned the compensation system. These officers asserted that they were not serving for purely financial reasons, knowing that they are serving for what many of them described as "25 percent of basic pay"—that is, they are already eligible for retirement pay of 75 percent of their current basic pay. Shortcomings of the current compensation system will become more evident if officers serve for longer careers. The compensation is capped, and continued service will have even more considerable opportunity costs. Further, longer service reduces the likelihood and/or length of meaningful post-service employment and should not exact a penalty for lifetime income streams. Our recommendations in the next chapter address this issue further.

ADDITIONAL OBSERVATIONS FROM THE INTERVIEWS

Several other issues of interest emerged from the interviews:

- **The current culture is one of mobility and movement.** Currently, accession to G/FO rank is perceived by officers as the entry to the exit ramp and not a time of broadening themselves in preparation for providing maximum contribution to organizations. This could be a result of the culture, which is described as one of mobility and movement: The military goes somewhere to do business, and movement has thus become a way of life with officers and families—you serve at the pleasure of the chief of service.

- **The current system allows for performance changes with grace.** When there are performance problems, the current system allows for a graceful exit because people do routinely retire. A changed system with longer tenures will need to deal with performance issues more directly.

- **Some assignments may resist lengthening.** There needs to be a recognition that some jobs are lacking in terms of perks, relationships, time, location, and satisfaction from seeing accomplishments. Moreover, some jobs are too demanding to be done for long periods.

- **If there are to be changes, the service chiefs should be the longest O-10 assignments.** There is a learning curve to the job and a high level of internal and external pressures; the officers need time to make changes. Combatant commanders are "here and now" people. Longer service than now is better, but they may need to change more frequently based on geographic and warfighting expertise needed at any given time. In general, certain kinds of assignments should be longer, while assignments to operational positions should be about the same as they are now, particularly those that develop officers; this is another reason that flexibility in the system is important.

- **The emerging policy of advancing deputies and vices to the head of the organization may not be the best idea in all situations.** Frequently in this age, the vices have agendas and responsibilities of their own to work for the organization and may need a unique skill set to do it. Having them as clones of the lead because of planned succession might not always be the best solution. Each situation will be unique.

- **Expectations of officers and organizations need to be set.** The first or second assignment determines the future of the officer. Experiences and "quality" of the tour matter as much as length for continued promotability. Officers need to be told where they stand; organizations need to know what tenure they can expect from their leadership.

- **Need to balance whole system: management, compensation, retirement, working conditions.** These parts are looked at periodically, but a comprehensive solution is needed. Generally, the high performers are not working for money, but compensation does matter. There needs to be succession planning shared with the officer and mentors. The Chairman needs to be part of succession planning along with the service chiefs. The civilian leadership has to maintain its ability to change military leadership. Turbulence and frenetic working conditions caused by frequent turnover can be reduced.

Chapter Seven

CONCLUSIONS AND RECOMMENDATIONS

CONCLUSIONS

With the exception of some O-10 jobs (e.g., chief of service, Chairman of the Joint Chiefs of Staff), the current management system generally does not determine assignment tenure based on either the inherent qualities of different positions or the way these assignments are used to develop officers. Some assignments should be longer than others. By making the distinction between developing jobs and using jobs, the length of some assignments can be extended—which mitigates the Secretary of Defense's concerns—without blocking the promotion of officers—which is a concern of the military services. These longer assignments can coexist with equal or better throughput and promotion probability, although some decreased time in career in the O-7 pay grade results. Thus, it is possible to extend assignments for the most-senior officers and for some selected O-8 and O-9 assignments without limiting the developmental opportunities for "fast-trackers" destined for further promotion.

The analysis described herein suggests the value of a revised system based on the career model described previously. Such a revised system can increase individual accountability, contribution to organizations, and overall organizational stability and, as a result, increase organizational performance. Time in job is managed, and career tenure and time in grade at retirement become second-order outcomes. Moreover, such a system clarifies expectations for the officers in it.

Currently, officer management is generally thought of in terms of "managed flow," and its language is that of fluid dynamics.[1] A central questions is: Where do officers go next, and how fast? While more attention is paid to the career development of G/FOs, including consideration of which positions will best prepare officers for senior leadership assignments and which individuals are the best candidates for current and future opportunities, the officer management system overall has been focused on rapid movement, as our data show. The metrics are time spent in grade, time spent in service, and promotion timing and promotion probability. The development rule is that all flow on a fixed schedule, with a few exceptions in which the Secretary and the chief of service believe more time is needed to develop skills and have desired organizational impact. The private sector uses comparable processes in terms of frequent movement of high-potential executives to positions of higher responsibility as a means to test and develop them and for the organization to learn about their capabilities for higher positions. However, the private sector slows the movement of individuals at the apex of their careers, even if they have not reached the highest levels of management. Further, because the ends of this process are to find the best people for such terminus positions as CEO, service in these positions is longer and continues until performance expectations are not met, an individual leaves voluntarily, or a company age-based retirement point is reached. This is a "managed use" system. With managed use, the central question is: Where do they come from, and how good are they for higher-level positions? The metrics are performance, the "quality" of past experience, and the quantity of it.[2] You do not necessarily become broad by the accumulation of specialties but rather by seeing the business as a whole. Retention is governed by performance, and in development, some move and some do not.

---

[1]See, for example, "Part III: Officer Flow Data," in DoD (2000). This section of the *Defense Manpower Requirements Report* provides Flow Management Plans for each of the military services for the commissioned officer grades of O-1 to O-10. The health of the officer management system is demonstrated by flows: the promotes "in" and "out" over a five-year period in response to retirements and other losses.

[2]"Achieving generalization by accumulating specialties is not sufficient for obtaining the broad view of the organization necessary for high-level management effectiveness" (John Boon, RAND, unpublished research). Also see Drucker (1954).

While the military services do consider metrics of the type discussed above, their major emphasis is on movement. There are consequences for the military if such a managed use system were more fully adopted. First, departures would be more episodic than routine. Upon selection of a long-serving executive, those who are competitive but not selected would be expected to leave for other opportunities. Departures cluster around successions. Second, performance departures would be more visible because they would occur with much less service than expected. As one officer stated, the system would be less graceful. If the private-sector experience is replicated (and recognizing a chicken-and-egg proposition), longer tenure in terminus positions would be correlated with greater organizational performance.

While we have focused so far on certain line officers and their paths to O-10, the logic of developing and using positions should also apply to the other communities that culminate in pay grades below O-10. Indeed, even longer using assignments may be appropriate in some of the professional occupations within the military (e.g., doctors), and our interviews suggest that changing tenures for these would be more easily accepted. One policy prescription (which would require a change to Title 10) might be to do away with the intermediate tenures that exist for O-7 to O-10 and let only the waiverable age retirement point work within a managed use system. Are there impacts from doing this? Yes, but our modeling suggests that there are means to dissipate such effects.

We conclude that implementation of the management changes described above could be done largely without additional legislative authority. Title 10 authority for 40-year careers for O-10, 38-year careers for O-9, and retirement at age 62 (along with the current waiver authority in law) generally is sufficient. However, a change in law could simplify the management of G/FOs while providing needed flexibility. For example, it should be possible to extend the assignments of chiefs of service, the Chairman, and the other officers with fixed terms, absent the condition of war or national emergency, which is required under existing law. Allowing officers to retire with less than three years time in grade would allow them to leave as

needed.[3] Also, the normal (that is, Social Security–eligible) retirement age in the United States is being extended to 67, and there
appears to be no reason why the military retirement age should not
increase as well. (The current age of 62 in law dates from the 1860s;
150 years have seen great strides in health, vigor, and longevity.)

Although this system could mostly be implemented within current
legislative management constraints, DoD should consider requesting
compensation changes. Such changes could include "uncapping"
pay at senior levels, continuing the accumulation of retirement ben-
efits to 100 percent at 40 years of service, and basing retirement pay
on uncapped figures. Additionally, because some officers who serve
in the shorter developing assignments will not be promoted, they will
require time-in-grade waivers at retirement. High-3 retirement will
be an issue for these officers in future years because it bases retirement pay on the highest three years of basic pay, and they will not
have spent three years in their last pay grade.[4]

Finally, such a changed system will require some flexibility. For
example, performance shortcomings will need to be dealt with
promptly and directly because longer assignments are not conducive
to continuing a nonperformer until retirement. Just as approximately
15 percent of CEOs are terminated for performance reasons, the military should anticipate a small number of officers who will require
separation prior to completion of a longer assignment.

RECOMMENDATIONS

We consider the input from interviews in our recommendations, but
the recommendations are largely drawn from the analysis. To
implement a changed system, the services should identify positions
as either developing or using and then set goals for desired tenure in
a position. Each service has accomplished or has under way a review
of the occupational experience and skills needed for successful

---

[3]At the time of this report, Congress was considering this as part of the National
Defense Authorization Act for Fiscal Year 2004.

[4]The magnitude of the dollar effect is dependent on annual pay adjustments, usually
based on the employment cost index, and any future structural adjustments to the pay
table.

performance in various G/FO positions. Job responsibilities, as well as desired developmental and experience assessments for high-potential officers, should form the basis of the position designation.

We stress for several reasons that our observed using–developing splits for each of the services should not be used prospectively for management without review. First, it needs to be confirmed that the services are developing officers with the assignments they historically have used to do this. Second, we acknowledge that some developing jobs may be better as three-year jobs and that some using jobs are not appropriate for a four-year tenure. This may especially be the case in overseas positions or particularly taxing jobs. Moreover, no system should be too rigid. For example, officers could be promoted after completing a using assignment, and operational using assignments at the O-10 level may not always be as long as other using assignments.

The optimum time in job should vary by pay grade, community, and the nature of the position. Thus, using assignments would be longer than developing assignments. Ideally, developing assignments for line officers would be a minimum of two years and using assignments a minimum of four years. Jobs for those outside the line community may also be longer than those within the line. Such a revised system would place the emphasis on managing to time in job and allowing time in grade or time in service to adjust to improved time in job. Additionally, the services should manage the numbers of developing assignments that officers have at grades O-7 and O-8. Given the developing and using split empirically observed in our modeling, three developing jobs overall at O-7 and O-8 and one developing job at O-9 maximize the amount of development while remaining feasible. More developing assignments at O-7 and O-8 become infeasible in that there are not enough officers available for promotion to O-9. More jobs may be possible if officers serve longer at O-9, or if the developing and using split identified by the services is significantly different from that used in our modeling.

This research proposes a system with greater stability and accountability, with fewer job rotations and longer service in position for many. However, we recognize that the military culture is one of mobility and movement, and transition to a new career model will encounter hurdles. We do not anticipate overall retention problems.

Our interviews with senior leaders, while not representative, suggest that most would stay longer if asked to do so and that retention will likely continue to be an individual issue, related to family concerns and other factors. Regardless, the services will need fewer officers to enter into the highest levels in the new system, so there is room for some officers to decline the longer tenure and leave, because the pool from which to pick the most-senior officers is large compared with the number selected.

DETAILED ANALYSIS OF CURRENT MANAGEMENT OF GENERAL AND FLAG OFFICERS

This appendix includes more-detailed data for assignment length, time in grade, time in service, and time to O-7 promotion in the current general and flag officer (G/FO) system. The data reflect Army infantry, armor, and artillery officers; Navy unrestricted line officers; Air Force pilots and navigators; and Marine Corps line officers.

LENGTH OF ASSIGNMENTS[1]

Figures A.1–A.4 graph the average tenure in assignment. Each figure compares time in job across grades but within services. As will be seen in the next few figures, it is unlikely for an assignment to be from 36 to 48 months or longer. In this analysis, median lengths provide the best empirical portrayal.

Figure A.1 reveals that the Army O-7 assignment length distribution is bimodal, with times concentrating around 12 and 24 months and a median of 15 months. Army O-8 assignments peak at 24 months with a median of 23, but contain a noticeable blip at 36 months; O-9 and O-10 assignments are widely distributed over 12 to 36 months with medians of 26 and 27, respectively.

---

[1]The data for Figures A.1–A.8 were derived from the General and Flag Officer database provided by DIOR. Figures A.1 to A.4 show length of assignments for all assignments that began on or after January 1, 1990. Figures A.5 through A.9 show time in grade for all officers promoted to that grade on or after January 1, 1990, and have either retired from that grade or been promoted to the next grade.

Figure A.2 shows that Navy assignment lengths follow the same pattern as those found for Army assignments, except for higher variation at all grades and a less noticeable bimodal distribution at O-7 resulting in an O-7 median five months higher.

Very similar to the Navy pattern is the Air Force pattern in Figure A.3. For Air Force O-7 assignment length, a peak at 24 months just barely rises from the almost-uniform density from 12 to 24 months. The median O-8, O-9, and O-10 assignment lengths are 20, 23, and 24 months, respectively, which are a few months shorter than the Navy's 22, 25, and 28 months.

The Marine Corps assignment length pattern in Figure A.4 is consistent with that of the Army. An underlying bimodal 12 and 24 months can be distinguished for O-7s, O-8s, and O-9s, although the small sample size of O-10s precludes distributional conclusions. From the perspective of planning and executing time in assignments, the Marine Corps appears to be the most "tightly" managed G/FO system. Time in job is far more closely centered around the 12 and 24 month modes for O-7s, O-8s, and O-9s, while O-10s center around 12, 24, 36, and 48 months. This reduced "noise" cannot be explained by a sample size one-fifth that of the other services and is likely a result of structural elements in the Marine Corps' general officer system.

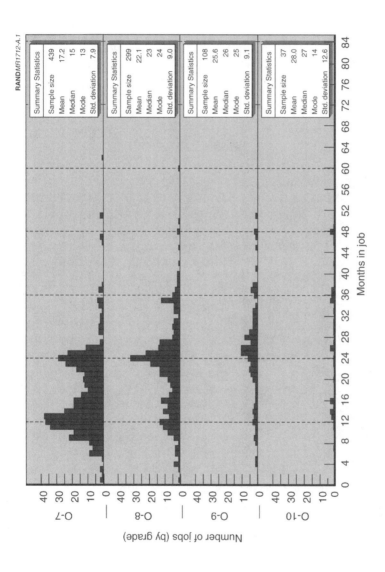

RAND*MR1712-A.1*

Summary Statistics	
Sample size	439
Mean	17.2
Median	15
Mode	13
Std. deviation	7.9

Summary Statistics	
Sample size	299
Mean	22.1
Median	23
Mode	24
Std. deviation	9.0

Summary Statistics	
Sample size	108
Mean	25.6
Median	26
Mode	25
Std. deviation	9.1

Summary Statistics	
Sample size	37
Mean	28.0
Median	27
Mode	14
Std. deviation	12.6

Months in job

Number of jobs (by grade)

O-7   O-8   O-9   O-10

Figure A.1—Army Time in Job: 1990 to June 2002

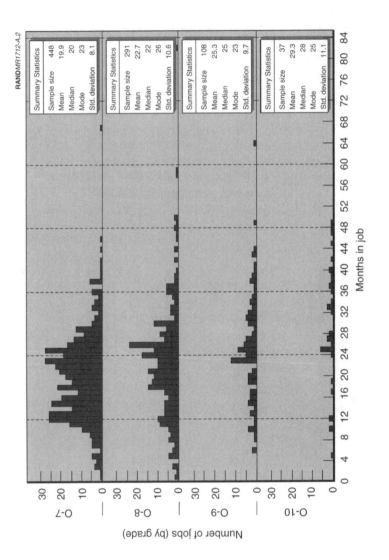

Figure A.2—Navy Time in Job: 1990 to June 2002

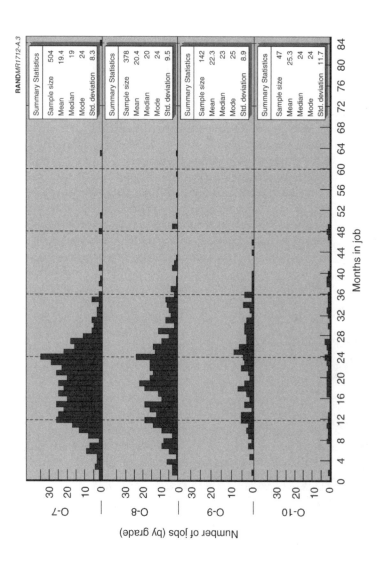

Figure A.3—Air Force Time in Job: 1990 to June 2002

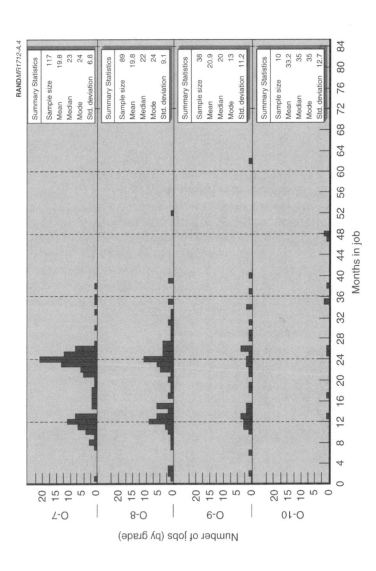

Figure A.4—Marine Corps Time in Job: 1990 to June 2002

TIME IN GRADE

In Figures A.5–A.8, we examine in the aggregate the time that G/FOs have spent in grade. The height of the bars in each of the following figures can be read as the number of officers for each service and pay grade who served a given number of months in one grade before retiring or being promoted to the next.

Figure A.5 shows that 71.3 percent of Army O-7s serve between 30 and 42 months in grade, with a minority serving 48 months or longer and a median of 38.5 months. O-8, O-9, and O-10 times are further spread out: O-8s range from three to 95 months, O-9s from 11 to 68 months, and O-10s are scattered from 14 to 97 months. Their medians are 36, 30, and 47 months, respectively.

Figure A.6 shows a similar pattern for Navy time in grade. 65.7 percent of Navy O-7s serve between 30 and 42 months in grade. O-7 time in grade peaks at 34 months and has a more diverse distribution and a lower median of 35 months. The O-8, O-9, and O-10 distributions are flat across similar ranges, with medians of 37.5, 33.5, and 35.0 months, respectively. However, the O-10 high is an extreme of 100 months.

The Air Force O-7 distribution in Figure A.7 is centered around 36 months, with a median of 37 months, with 30–42 months having 61.4 percent of all observations. Air Force O-8s, O-9s, and O-10s present virtually identical patterns to both the Army and the Air Force distributions, except Air Force O-8 has three visible modes at 24, 36, and 48 months. The medians are 37, 33, and 39 months for O-8s, O-9s, and O-10s, respectively.

Once again, the smaller number of observations of the Marine Corps data makes comparison both more difficult and more fruitful. 50.8 percent of Marine Corps O-7s serve between 30 and 42 months. For all grades, modes can be seen at 24, 36, and 48 months, with little "noise" in between these ranges. O-8 has an additional mode at 12 months and additional observations around 60 months. However, the medians, 36, 35, 34, and 36 months, are more centered at three years per grade than in any other service.

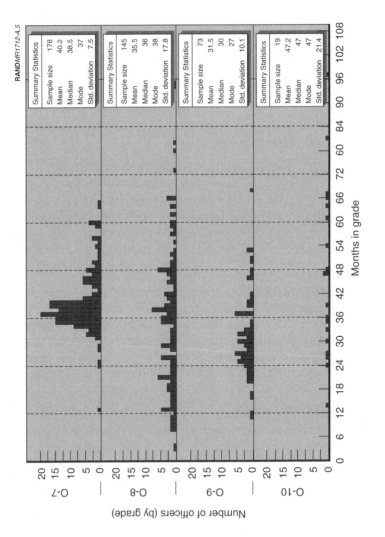

Figure A.5—Army Time in Grade: 1990 to June 2002

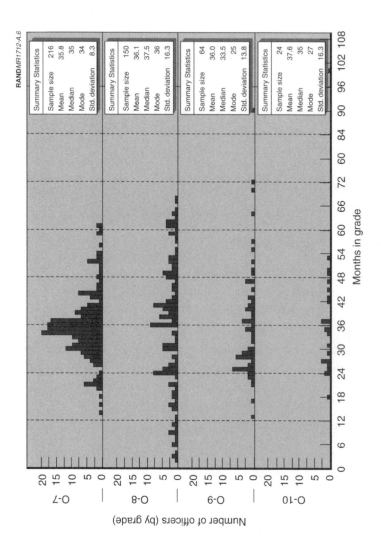

Figure A.6—Navy Time in Grade: 1990 to June 2002

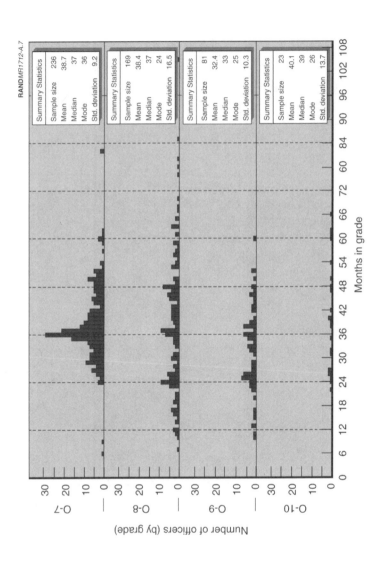

Figure A.7—Air Force Time in Grade: 1990 to June 2002

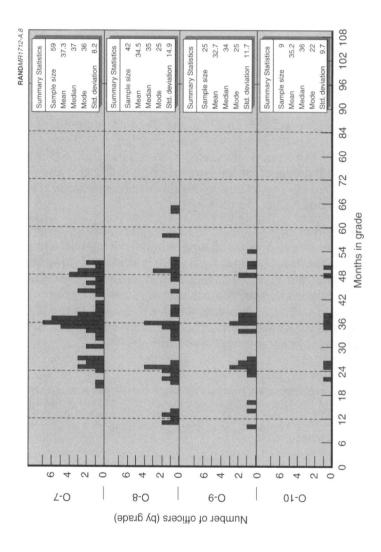

Figure A.8—Marine Corps Time in Grade: 1990 to June 2002

TIME IN SERVICE AT RETIREMENT[2]

Figures A.9–A.12 look at the career lengths of G/FOs. The vertical height of the bars indicates the number of officers for each service who retired at a given pay grade with a specific number of years of service.

The patterns in Figure A.9 are typical for all four services; with the exception of Marine Corps O-8s, all four services at all pay grades have single-mode distributions of time in service. Time in service is a few years lower for O-7 than for O-8, O-9, and O-10, but the differences between O-8 and O-9 and O-9 and O-10 are generally much smaller. Specifically, the Army's O-7 has a median of 30.2 years, but its O-8, O-9, and O-10 have medians of 33.2, 34.1, and 35.3 years, respectively. The medians for the Navy are 31.3, 34.1, 33.9, and 35.0 years; 29.1, 31.7, 34.0, and 34.3 years for the Air Force; and 30.2, 34.2, 34.2, and 36.2 years for the Marine Corps.

In all services, retiring O-10s serve longer than retiring O-9s, but retiring O-9s do not always serve longer than retiring O-8s. In the Army, retiring O-9s have 0.9 more years of service than retiring O-8s. In the Air Force, retiring O-8s have 2.3 more years of service than retiring O-8s. However, in the Marine Corps, retiring O-9s serve the same amount of years of service as retiring O-8s, and in the Navy, retiring O-9s serve 0.2 fewer years of service than retiring O-8s. In all services, retiring O-8s serve longer than retiring O-7s.

---

[2]The data for Figures A.9 through A.16 are for officers in line communities who retired after January 1, 1990. The data were generated by merging the General and Flag Officer database from DIOR with the Joint Duty Assignment Management Information System from the Defense Manpower Data Center. Observations with incomplete or corrupt data were removed.

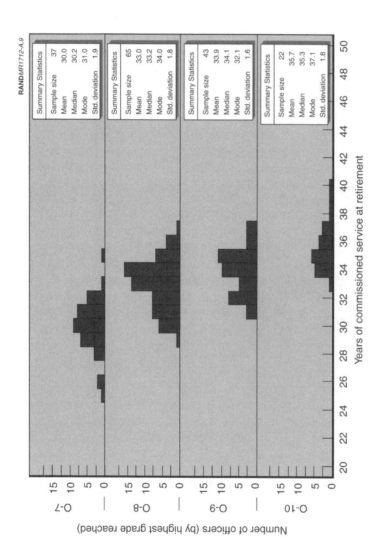

Figure A.9—Army Time in Service: 1990 to June 2002

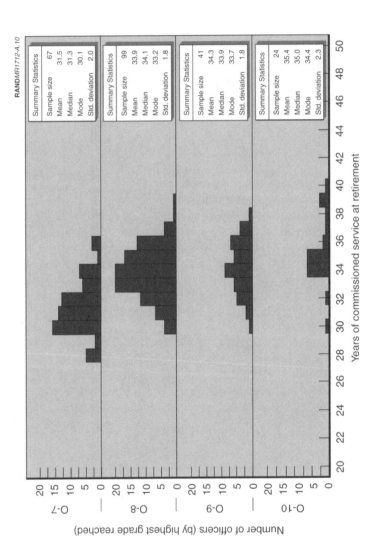

Figure A.10—Navy Time in Service: 1990 to June 2002

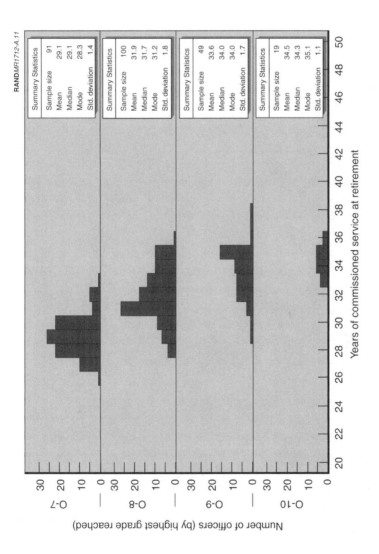

RAND*MR1712-A.11*

Summary Statistics

	O-7	O-8	O-9	O-10
Sample size	91	100	49	19
Mean	29.1	31.9	33.6	34.5
Median	29.1	31.7	34.0	34.3
Mode	28.3	31.2	34.0	35.1
Std. deviation	1.4	1.8	1.7	1.1

Years of commissioned service at retirement

Number of officers (by highest grade reached)

Figure A.11—Air Force Time in Service: 1990 to June 2002

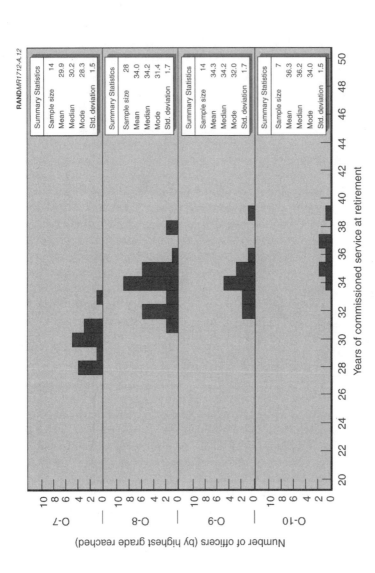

Figure A.12—Marine Corps Time in Service: 1990 to June 2002

TIME TO O-7

The next set of figures, examining the number of years it takes before officers are promoted to O-7 compared with the highest grade they will achieve before retiring, provides explanation for the apparent inconsistency between the much larger number of years that O-10s spend as G/FOs and the relatively similar amount of total years of service for all G/FOs, regardless of the pay grade from which they retired. As a general rule, eventual O-10s make O-7 earlier than O-9s, who make O-7 earlier than eventual O-8s, who make O-7 earlier than officers who do not get promoted beyond O-7.

Figure A.13 shows that for the Army, promotion to O-7 occurs with a median of 26.8 years of service for those not promoted beyond O-7, with a median of 25.8 years of service for retiring O-8s, a median of 25.1 years of service for retiring O-9s, and a median of 23.9 years of service for those eventually promoted to O-10. The standard deviation remains relatively constant over all four retiring pay grades.

Figure A.14 shows that for the Navy, promotion to O-7 occurs with a median of 27.7 years of service for those not promoted beyond O-7, with a median of 27.2 years of service for those who retire as O-8s, a median of 26.6 years of service for those retiring as O-9s, and a median of 25.7 years of service for those eventually promoted to O-10. Again, the standard deviation remains relatively constant over all four retiring pay grades.

Figure A.15 shows that for the Air Force, promotion to O-7 occurs with a median of 25.2 years of service for those not promoted beyond O-7, with a median of 24.9 years of service for retiring O-8s, a median of 24.6 years of service for retiring O-9s, and a median of 23.8 years of service for those eventually promoted to O-10. The standard deviation remains relatively constant over all four retiring pay grades but is consistently much smaller for the Air Force than for the Army and Navy.

Figure A.16 shows that for the Marine Corps, promotion to O-7 occurs with a median of 26.7 years of service for those not promoted beyond O-7, with a median of 27.5 years of service for retired O-8s, a median of 26.8 years of service for retired O-9s, and a median of 26.4 years of service for those eventually promoted to O-10. However, because of the much smaller sample size, comparisons are difficult to make with the other services. Standard deviations appear to correlate positively with the sample sizes, meaning that, the more officers in the sample, the greater the chance that more-extreme observations will vastly increase the measure of variation.

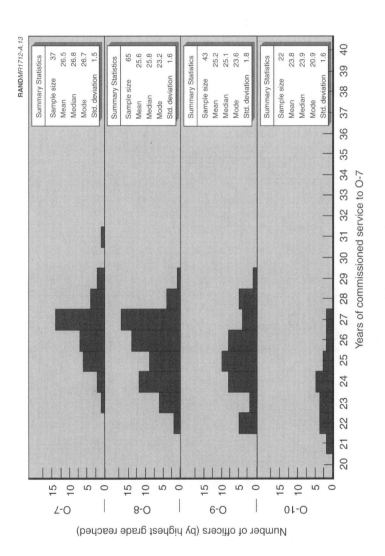

Figure A.13—Army Time to O-7: 1990 to June 2002

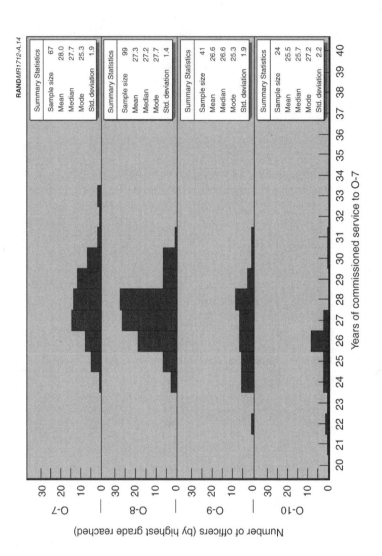

Figure A.14—Navy Time to O-7: 1990 to June 2002

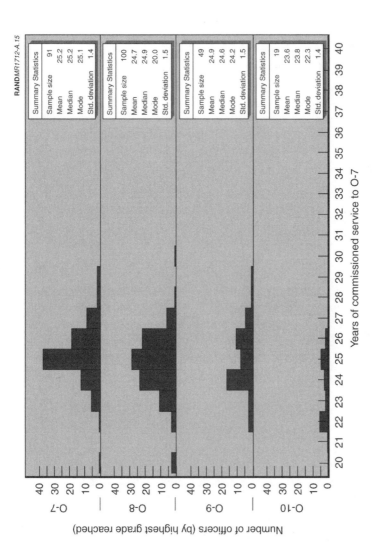

Figure A.15—Air Force Time to O-7: 1990 to June 2002

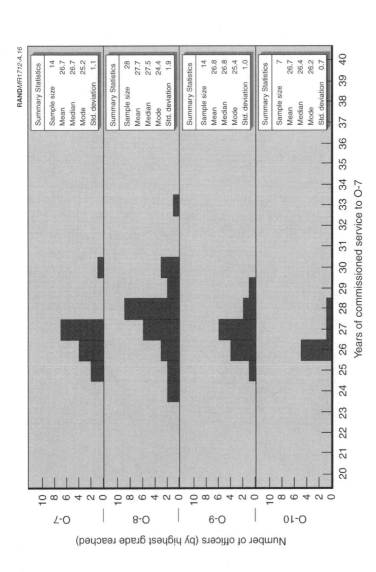

Figure A.16—Marine Corps Time to O-7: 1990 to June 2002

MODELED POLICY EXCURSIONS

Table B.1 lists the excursions that we modeled for each of the four services. The left column indicates the case number. Some cases use both letters and numbers, indicating a subsequent variation on a case presented previously to the sponsor office. The next two columns represent the length of the developing and using assignments in these excursions. The length occasionally varied by grade (e.g., longer using assignments for O-10s in cases 4 and 4a). Some cases explored the appeal of retaining developing officers who were not promoted for an additional year to provide them with three years time-in-grade promotion. In these instances, the length of developing assignments is noted as "2 + 1." The final two columns indicate the number of each type of job, per grade, that officers were assigned. For example, officers in case 2 received either two developing jobs or one using job at each pay grade. Again, there were cases in which the number of each type of job varied by pay grade. In 7d, officers received a total of three developing jobs during the time they were O-7s and O-8s. Thus, some officers might have one developing assignment while they were O-7s and two while they were O-8s, while others might have two developing jobs at O-7 before having a single developing job at O-8.

Case 7d is the career model presented in the body of this report.

Table B.1

Modeled Excursions

	Length of Assignment (years)		Number of Assignments (per grade)	
Case	Developing	Using	Developing	Using
1	2	4	1	1
2	2	4	2	1
3	2 + 1	4	1	1
4	2 + 1	4 for O-8, O-9; 6 for O-10	1	1
4a	2	4 for O-8, O-9; 6 for O-10	1	1
5	2 + 1	4	1	1 at O-8, O-9; 2 at O-10
5a	2	4	1	1 at O-8, O-9; 2 at O-10
6	2 + 1	4	1	1 at O-8; 2 at O-9, O-10
6a	2	4	1	1 at O-8; 2 at O-9, O-10
7a	2	4	2 at O-7; 1 at O-8, O-9	1 at O-8, O-9; 2 at O-10
7b	2	4	2 at O-8; 1 at O-7, O-9	1 at O-8, O-9; 2 at O-10
7c	2	4	3 at O-7/O-8; 1 at O-9	1 at O-8, O-9; 2 at O-10
7d	2	4	3 at O-7/O-8; 1 at O-9	1 at O-8; 2 at O-9, O-10

MODELING RESULTS IN TABULAR FORM

The detailed service-specific results of this system, for the previously identified subsets of officers (e.g., Navy unrestricted line, Air Force pilots and navigators), compared with the status quo for those population subsets, appear in Tables C.1 to C.4.

The tables indicate that the Army and the Navy will experience a greater number of officers promoted to O-7 annually; the Air Force will stay about the same as it is now; and the Marine Corps will see a slight decrease in new O-7s. This is related to the slightly shorter time in grade for O-7s, who are either promoted onward or leave after one or two two-year assignments. While they may spend less time in grade than in the status quo, their assignments are longer than those currently experienced by O-7s. All the services will see approximately equal or greater annual numbers of promotions to O-8. However, promotion probability to O-8 increases for three of the services, and decreases for one (the Army). Promotion probability to O-9 increases for three of the services (all but the Army), and as many as 90 percent (for the Air Force and the Navy) of officers in O-8 developing assignments will be promoted to O-9. O-9s who serve in using positions will remain in grade for eight years and retire with 39–41 years of military service, which is about six years longer than current O-9s serve before retirement. However, O-9s destined to be O-10s will move relatively quickly through grade O-9, serving one two-year assignment. Some of these developing O-9s will not be selected for promotion, and their average time in service will be similar to that of current O-9s. Thus, the average time in service for all retiring O-9s is about the same as today's overall average but with two distinct groups with different averages. O-10s serve in grade for eight years and retire with approximately 40–43 years of service, which is five or six years longer than in the status quo.

In sum, three of the services will experience a greater number of officers promoted to O-7 annually—the Marine Corps will see no change. Equal or greater numbers of officers will also reach grade O-8. Fewer officers will rise to O-9 and O-10 because the length of time that O-9s and O-10s serve before retirement will increase considerably. Average career length will increase for all pay grades except O-7; however, O-7s will spend longer time in assignments than is experienced in the status quo.

Although this system addresses the shortcomings perceived in the status quo by lengthening selected assignments and extending the time that O-8s, O-9s, and O-10s spend in their final grade before retirement, these changes will not necessarily be evident in the metrics currently used to monitor the system. For example, an average time-in-grade calculation that includes those promoted out of the pay grade as well as those retiring from that grade would not indicate much change because promotees will likely have served in two-year developing assignments. Additionally, those nominated for promotion will still show quick movement through the system, given their resume of developing assignments. Metrics that track the dual populations or developers and users separately, such as time in grade at retirement or promotion probability, will prove more useful.

Table C.1

Comparison of New Results with Status Quo:
Army Infantry, Armor, Artillery

	Grade			
	O-7	O-8	O-9	O-10
Current Practice				
Number promoted	20.8	17.8	8.4	2.4
Promotion probability	—	86%	47%	28%
Average assignment length (months)	17.2	22.1	25.6	28.0
Average time in grade at retirement (years)	3.7	3.7	2.7	3.9
Average career length (years)	30.0	33.0	33.9	35.7
Alternative				
Number promoted	22.3	18	7.1	1.3
Promotion probability	—	81%	39%	18%
Promotion probability—developers	—	81%	59%	24%
Promotion probability—users	—	N/A	0%	0%
Developing jobs/using jobs	100%/0%	60%/40%	45%/55%	0%/100%
Average assignment length (months)	24.0	33.6	37.2	48.0
Average time in grade at retirement (years)	2.5	3.2	2.6	8
Average time in grade at retirement—developers (years)	2.5	2.2	2	N/A
Average time in grade at retirement—users (years)	N/A	4	8	8
Average career length (years)	29.0	31.8	35.6	40.4
Average career length—developers (years)	29.0	30.8	33.9	N/A
Average career length—users (years)	N/A	32.6	39.9	40.4

NOTE: N/A = not applicable.

Table C.2

Comparison of New Results with Status Quo: Navy Unrestricted Line

	Grade			
	O-7	O-8	O-9	O-10
Current Practice				
Number promoted	24.6	17.1	7.0	2.8
Promotion probability	—	70%	41%	39%
Average assignment length (months)	19.9	22.7	25.3	29.3
Average time in grade at retirement (years)	3.5	3.6	3.1	3.1
Average career length (years)	31.5	33.9	34.3	35.4
Alternative				
Number promoted	26	18.4	8.0	1.3
Promotion probability	—	71%	44%	16%
Promotion probability— developers	—	71%	90%	27%
Promotion probability— users	—	N/A	0%	0%
Developing jobs/using jobs	100%/0%	35%/65%	40%/60%	0%/100%
Average assignment length (months)	24.0	39.6	38.4	48.0
Average time in grade at retirement (years)	2.2	4.0	2.7	8
Average time in grade at retirement—developers (years)	2.2	3.4	2	N/A
Average time in grade at retirement—users (years)	N/A	4	8	8
Average career length (years)	29.9	34.5	35.4	41.4
Average career length— developers (years)	29.86	33.74	34.15	N/A
Average career length— users (years)	N/A	34.6	40.2	41.4

NOTE: N/A = not applicable.

Table C. 3

Comparison of New Results with Status Quo: Air Force Pilots and Navigators

	Grade			
	O-7	O-8	O-9	O-10
Current Practice				
Number promoted	26.6	18.8	8.0	3.1
Promotion probability	—	70%	43%	39%
Average assignment length (months)	19.4	20.4	22.3	25.3
Average time in grade at retirement (years)	3.6	3.7	2.7	3.3
Average career length (years)	29.1	31.9	33.6	34.5
Alternative				
Number promoted	27	20.5	11.5	1.3
Promotion probability	—	76%	56%	11%
Promotion probability— developers	—	76%	89%	12%
Promotion probability— users	—	N/A	0%	0%
Developing jobs/using jobs	100%/0%	56%/44%	72%/28%	0%/100%
Average assignment length (months)	24.0	34.6	30.7	48.0
Average time in grade at retirement (years)	2.3	3.8	2.2	8
Average time in grade at retirement—developers (years)	2.3	2.7	2	N/A
Average time in grade at retirement—users (years)	N/A	4	8	8
Average career length (years)	27.4	31.6	33.7	39.7
Average career length— developers (years)	27.4	30.5	33.2	N/A
Average career length— users (years)	N/A	31.8	39.2	39.7

NOTE: N/A = not applicable.

Table C.4

Comparison of New Results with Status Quo:
Marine Corps Line

	Grade			
	O-7	O-8	O-9	O-10
Current Practice				
Number promoted	7.1	5.0	3.0	1.1
Promotion probability	—	70%	60%	38%
Average assignment length (months)	19.8	19.8	20.9	33.2
Average time in grade at retirement (years)	3.2	3.5	2.8	2.9
Average career length (years)	29.9	34.0	34.3	36.3
Alternative				
Number promoted	6.0	5.0	3.2	0.4
Promotion probability	—	82%	65%	11.7%
Promotion probability—developers	—	82%	86%	12%
Promotion probability—users	—	N/A	0%	0%
Developing jobs/using jobs	100%/0%	70%/30%	89%/11%	0%/100%
Average assignment length (months)	24.0	30.0	26.6	48.0
Average time in grade at retirement (years)	2.4	3.2	2.1	8
Average time in grade at retirement—developers (years)	2.4	2.3	2	N/A
Average time in grade at retirement—users (years)	N/A	4	8	8
Average career length (years)	29.2	34.2	35.3	42.9
Average career length—developers (years)	29.2	33.1	35.1	N/A
Average career length—users (years)	N/A	34.6	41.1	42.9

NOTE: N/A = not applicable.

MODELING RESULTS IN "FLOW" FORM

The primary model used in this analysis is a "stock and flow" model, constructed in system dynamics software. The stock and flow nature of the model makes it relatively easy to portray the movement of officers through the model. This appendix provides the modeling results in flow diagrams to complement the tables included in Appendix C. Thus, Figures D.1–D.4 portray the modeling results of the final case for each of the four services. As an example, we discuss the Army case (portrayed in Figure D.1) below. Because the models are steady-state models, individuals are portrayed as fractions of numbers. Actual events would obviously not be in the form of fractional numbers, but for clarity of this explanation, the modeling results have been left in their fractional form.

The bottom-most row of stocks (boxes) reflects the steady-state movement of all officers through the general and flag ranks. Thus, 22.3 officers are promoted to grade O-7, and there are 67 Army O-7s in the occupations included.[1] Each year, 4.3 officers retire from the service as O-7s; they are shown leaving the system with a downward arrow. At the bottom is the account of what assignments these departing officers had served in (3.3 had served in one developing assignment; one officer left after serving in two developing assignments). The 4.3 departing-officers figure results from the difference between 22.3 annual promotions to O-7 and only 18 annual promotions to O-8. Said another way, of the 22.3 officers promoted annually

---

[1]As discussed earlier, this analysis included only those occupations that are promoted to O-10. This includes Army infantry, armor, field artillery; Navy unrestricted line officers; Air Force pilots and navigators; and Marine Corps line officers.

to O-7, 7.8 serve one developing assignment and are promoted to O-8, 10.2 serve two developing assignments before promotion to O-8, 3.3 retire after a single developing assignment, and 1.0 retires after serving two developing assignments.

Each year, 18 officers are promoted to grade O-8 (7.8 after one O-7 developing assignment, and 10.2 after two O-7 developing assignments). There are 60 O-8s in these communities. This equates to a promotion probability of 80.7 percent, computed as annual O-8 promotions (18) divided by 22.3 annual O-7 promotions. Approximately 11 O-8s retire each year, of whom 4.4 served one developing assignment, 0.5 completed two developing assignments, and six served in four-year using assignments. 7.1 officers are promoted to O-9 each year, equating to a promotion probability of 39.4 percent. 5.8 O-9s retire each year. Most O-9 retirees had served in developing assignments, and 1.7 served in a using assignment. Only 18.3 percent of O-9s will be promoted to O-10 (1.3 each year), and there are a total of 10 Army O-10s.

Above the main row of boxes are other stocks that represent the different kinds of assignments officers complete as they move through the system. These state the pay grade, whether they are developing or using assignments, and how many assignments they include (e.g., 24 O-8 using assignments). Where officers serve in multiple assignments, the stocks are broken out separately. Thus, as Army officers are promoted to O-7 each year, they are assigned to their first developing job. Of those 22.3 initial O-7s, 11.2 are assigned to a second O-7 developing job, 7.8 are promoted to O-8, and (shown at the bottom) 3.3 retire after the first developing assignment. Of the 11.2 O-7s who serve in a second O-7 developing assignment, 10.2 are promoted to O-8, and one retires.

Following Army officers through the O-8 pay grade in the figure, 12 officers are assigned to a first developing job. Of these, six complete a second O-8 developing assignment, 1.6 are promoted to O-9, and 4.4 depart after the first O-8 developing assignment. Of the six assigned to a second developing job, 5.5 are promoted, and 0.5 exit the system. Each year, six O-8s are assigned to using jobs, of which there are 24 total. These officers retire at the completion of their four-year assignment.

Of the 7.1 officers promoted to O-9, 5.4 are assigned to a developing job (1.3 of these officers will be promoted to O-10); the rest (1.7) O-9s are assigned to their first (of two) four-year using assignments. Officers promoted to O-10 will complete two four-year using assignments before retiring.

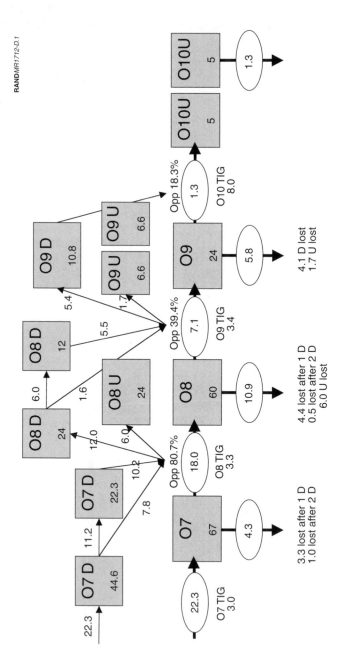

Figure D.1—Modeling Results in Flow Format: Army

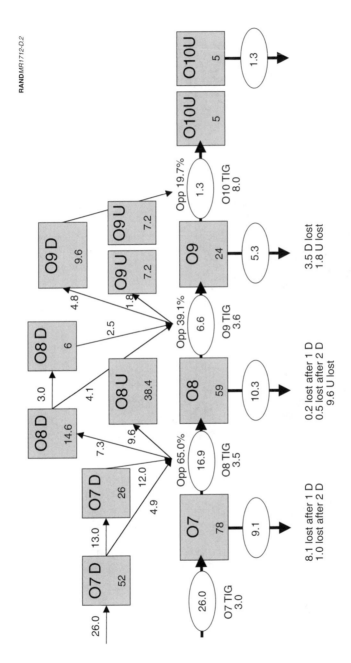

Figure D.2—Modeling Results in "Flow" Format: Navy

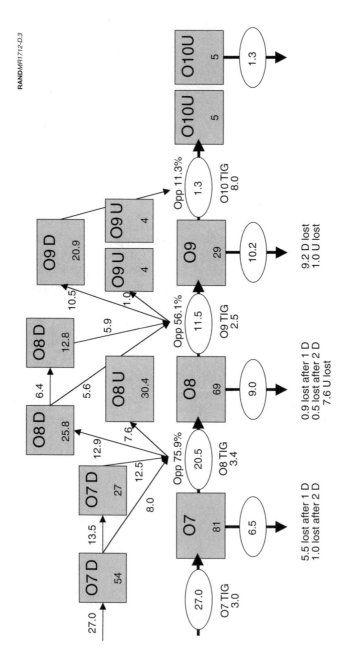

RAND*MR1712-D.3*

Figure D.3—Modeling Results in "Flow" Format: Air Force

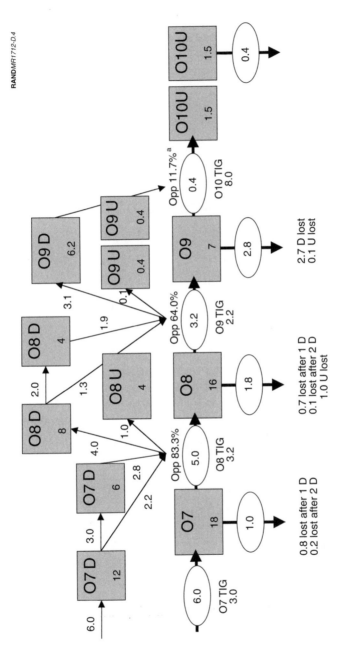

Figure D.4—Modeling Results in Flow Format: Marine Corps

[a]11.7 is the accurate promotion opportunity. The number of officers promoted to O-9 (3.2) and to O-10 (1.4) have been rounded and would thus erroneously imply a higher opportunity.

COMPENSATION OBSERVATIONS

This report shows the effects of a new career model on longer tenure in job and longer tenure in service. We compared tenure of O-10s with that of private-sector CEOs. This appendix offers observations on the compensation of the two groups. We do not suggest whether or not O-10 or CEO compensation is correct; rather, we contrast the two and offer observations from the research team.

Compensation of CEOs is in the public domain (through U.S. Securities and Exchange Commission filings) and has been illustrated in recent reports. For example:

- The top two executives at 187 Washington, D.C.–area companies have median total compensation between $1 million and $2 million.[1]

- CEOs of large corporations make 411 times as much as the average factory worker.[2]

- CEOs have a mean base salary of $366,000.[3]

- The median total direct compensation for CEOs of the 350 largest U.S. public companies is $6 million.[4]

---

[1] O'Hara (2002).

[2] "The Crisis in Corporate Governance" (2002).

[3] Hobgood (2002).

[4] Anders (2003).

Within a public corporation, CEO compensation is usually set by the compensation committee of the company's board of directors. This committee will typically review compensation consultant assessments that focus on industry, geographical, and market trends and examine four broad areas: salary, short-term incentives, long-term incentives, and perquisites, including special retirement plans. The committee is required to provide its own assessment, in the annual report (usually the 10k proxy statement), of executive pay and of CEO pay and show the data for the five highest-paid executives. The committee typically makes its decisions on pay based on such factors as company performance (absolute and relative to a peer group of companies), the individual's success in meeting goals set by the board, and the board's ability to retain and motivate the executive.

The compensation of O-10s is also in the public domain because it is contained in Title 37 of the U.S. Code:

- O-10s receive basic pay of about $147,900; the Chairman of the Joint Chiefs of Staff, the Vice Chairman of the Joint Chiefs of Staff, and chiefs of service receive basic pay of about $163,200. However, all are capped at Level III of the Executive Schedule, or about $142,500 annually. There are additional cash pays and cash and in-kind allowances and tax advantages that bring regular military compensation to approximately $175,500 annually. The last structural adjustment to basic pay is at 26 years of service. No additional basic pay is received for longer service.

- The Chairman of the Joint Chiefs of Staff's regular military compensation (capped) is less than six times the regular military compensation of a new recruit (E-1, which is about $30,300).

- The maximum retirement package is based on 30 years of service; it is 75 percent of capped basic pay, or approximately $106,900, and is subject to an annual cost-of-living adjustment.

The mechanism for O-10 pay setting is straightforward. The Executive Branch and Congress change pay and allowances, usually once a year, through an employment cost index adjustment or by other means.

We observe that:

- Private-sector executive pay is variable for each individual and rewards performances and accretion of human capital. CEO compensation is measured against market comparables (CEOs in similar industries), leadership and management abilities, and company performance.

- All O-10s are paid the same.

- Private-sector executive pay is "tilted up" in that it increases rapidly and significantly at the highest levels.

- O-10 pay is compressed and capped.

- Fewer officers will reach O-10 with a changed career model, and those who do will have longer periods of responsibility and accountability. No rewards exist for such a longer contribution of talent. In fact, penalties exist because O-10s forgo the 75 percent of basic pay at retirement for which they are eligible for up to 10 years.

- O-10 opportunity costs are high. Most O-9s and all O-10s will be working for 25 percent of their capped basic pay for nearly 10 years. Moreover, with later retirement, there is less opportunity to serve as a private-sector executive or director. Lifetime earnings streams are potentially reduced with longer service.

- O-10 retirement is based on one facet of pay—basic pay. At retirement, housing and other allowances are not considered. Retirement pay is 75 percent of capped basic pay. Moreover, the pay table ends at 26 years of service, so no structural adjustments to pay have occurred for up to 14 years beyond that point.

- Private-sector CEO retirement is more carefully managed and based on wealth accumulation and lifestyle maintenance. It includes continued company ownership (e.g., previously vested restricted stock), deferred compensation, consulting fees, perquisites, and special executive retirement plans (also known as "top hat" plans).

REFERENCES

Anders, George, "Upping the Ante," *Wall Street Journal,* June 25, 2003.

Campion, Michael A., Lisa Cheraskin, and Michael J. Stevens, "Career-Related Antecedents and Outcomes of Job Rotation," *Academy of Management Journal,* Vol. 37, No. 6, December 1994, p. 1535.

Conference Report to Accompany H.R. 3230, Washington, D.C.: U.S. Government Printing Office, 1996.

"The Crisis in Corporate Governance," *BusinessWeek,* May 6, 2002.

Derr, C. Brooklyn, Candace Jones, and Edmund L. Toomey, "Managing High-Potential Employees: Current Practices in Thirty-Three U.S. Corporations," *Human Resource Management,* Vol. 27, No. 3, Fall 1989, p. 275.

DoD—*see* U.S. Department of Defense.

Drucker, Peter R., *The Practice of Management,* New York: Harper & Row, 1954.

Forbes, J. Benjamin, and James E. Piercy, *Corporate Mobility and Paths to the Top: Studies for Human Resource and Management Development Specialists,* Westport, Conn.: Quorum Books, 1991.

Gabarro, John J., *The Dynamics of Taking Charge,* Boston: Harvard Business School Press, 1987.

Hadlock, Charles J., Scott Lee, and Robert Parrino, "Chief Executive Officer Careers in Regulated Environments: Evidence from Electric and Gas Utilities," *Journal of Law and Economics*, Vol. 45, October 2002, pp. 535–563.

Hobgood, Cynthia, "Highly Paid CEOs: Study Shows Where Buck Stops," *Washington Business Journal*, August 2, 2002.

Lucier, Chuck, Eric Spiegel, and Rob Schuyt, "Why CEOs Fall: The Causes and Consequences of Turnover at the Top," *Strategy & Business*, Third Quarter, 2002.

McCall, Morgan W., Jr., Michael M. Lombardo (contributor), and Ann M. Morrison (contributor), *The Lesson of Experience: How Successful Executives Develop on the Job*, New York: The Free Press, 1989.

Morrison, Robert F., and Roger R. Hock, "Career Building: Learning from Cumulative Work Experience," in Douglas T. Hall, ed., *Career Development in Organizations*, San Francisco: Jossey-Bass, 1986, pp. 237, 240–241.

Mowday, R. T., R. M. Steers, and L. W. Porter, "The Measurement of Organizational Commitment," *Journal of Vocational Behavior*, Vol. 14, 1979, pp. 223–247.

National Defense Authorization Act for Fiscal Year 1997.

O'Hara, Terence, "Washington Post 2002 Compensation Report: Study of 305 Executives at 187 Washington-Area Companies," *Washington Post*, July 22, 2002, p. E1.

U.S. Department of Defense, Under Secretary of Defense (Personnel and Readiness), *Defense Manpower Requirements Report*, May 2000.